Geochemistry

An Introduction

Francis Albarède

Ecole Normale Supérieure de Lyon
Institut Universitaire de France

Translated from French and expanded

English foreword by
Albrecht W. Hofmann

CAMBRIDGE
UNIVERSITY PRESS

PUBLISHED BY THE PRESS SYNDICATE OF THE UNIVERSITY OF CAMBRIDGE
The Pitt Building, Trumpington Street, Cambridge, United Kingdom

CAMBRIDGE UNIVERSITY PRESS
The Edinburgh Building, Cambridge CB2 2RU, UK
40 West 20th Street, New York, NY 10011-4211, USA
477 Williamstown Road, Port Melbourne, VIC 3207, Australia
Ruiz de Alarcón 13, 28014 Madrid, Spain
Dock House, The Waterfront, Cape Town 8001, South Africa

http://www.cambridge.org

First published 2003
Fourth printing 2006

Printed in the United Kingdom at the University Press, Cambridge

Typefaces Times New Roman 10/13 pt and Stone Sans   *System* LATEX 2_ε [TB]

A catalog record for this book is available from the British Library

ISBN 0 521 81468 5 hardback
ISBN 0 521 89148 5 paperback

Contents

Foreword to the English edition

Modern geochemistry is a discipline that pervades nearly all of Earth science, from measuring geological time through tracing the origin of magmas to unravelling the composition and evolution of continents, oceans and the mantle, all the way to the understanding of environmental changes. It is a comparatively young discipline that was initiated largely by Goldschmidt in the 1930s, but its modern development and phenomenal growth started only in the 1950s. Although there are many journals dedicated to geochemical research, there have been remarkably few general geochemical textbooks that cover more than a limited segment of the full scope of modern geochemistry. This is one reason why Francis Albarède's new book is most welcome. It is written by the author of the authoritative and widely acclaimed *Introduction to Geochemical Modeling* (Cambridge University Press, 1995), and it is intended as an undergraduate introductory course in geochemistry. Its scope is large, though not all-inclusive, concentrating on the inorganic chemistry of the condensed part of our planet. Although it started out as a translation from the original French book, the new English-language edition is much more than just a translation. The entire text has been substantially revised and in some parts expanded, and it is really a new book. Yet it retains a distinctly French flavor, particularly in the way many subjects are addressed via mathematical description. This approach is entirely normal for a student of the French Ecole Normale or a French university, but it will surprise many American teachers and students of geology alike. So if you are teaching or taking a course in "Rocks for Jocks," this book is not for you. But if you are interested in an introduction to modern geochemistry as a quantitative science, this book is definitely for you. Francis Albarède often uses a light touch, not taking the subject (or himself) excessively seriously, he uses refreshingly surprising analogs to approach important principles or processes, and his style is often informal. Look, for example, at the "Further reading" list. The books are classified into three categories, A, armchair reading, B, for students, and C, serious stuff, and each book is given a one- or two-line thumbnail characterization. Very nice. And by the way, the book itself, while clearly aimed at B, does contain material in all three categories!

What I particularly like about this book is its scope and choice of subjects, combined with sometimes bold brevity, which I hope will leave the student with an appetite for more. The emphasis is always on general principles rather than specific geochemical results or observations, and this should give the book a long residence time and keep it from becoming outdated. We are led from an introduction to the atomic and nuclear properties of the chemical elements to the principles of chemical and isotopic fractionation and mixing, geochronology and the use of radiogenic tracers to characterize source reservoirs, geochemical transport by advection and diffusion, the concept of closure temperature, chromatography, reaction rates to the treatment of large-scale systems, such as the oceans, the crust, and the mantle. The approach is initially mostly theoretical, focusing on the mathematical description of the behavior and interaction of single and multiple reservoirs. This is followed by a wide-ranging chapter on "Waters present and past," which covers topics from solution chemistry, water–rock interactions, erosion, rivers and oceans, to climate development during the Pleistocene. From there, we move to the "solid Earth," which deals with the evolution of mantle and crust, but also with the geochemistry of magmas. Finally, or almost finally, we are taken to phenomena of even much larger scale, the formation of the chemical elements in stars, the formation of the Solar System, the age and composition of Earth, Moon, and Mars.

The message the student should take from this is that geochemistry is a quantitative science that has made decisive contributions to the understanding of all these subjects. It has thus become one of the central disciplines of Earth science, a fact that is not always reflected in undergraduate curricula. This book should help to correct this common deficiency in the training of Earth scientists. It is an inspired book; I hope you will enjoy reading it as much as I did.

Albrecht W. Hofmann
Max Planck Institute for Chemistry, Mainz

Foreword to the French edition

I am specially happy to preface this book. First, because it is always a pleasure to be able to speak well of a friend's work; and Francis Albarède is a friend of long standing! We both embarked on our academic careers at about the same time. After some solid grounding in geology at the University of Montpellier, we were fortunate enough to begin our doctoral research in geochemistry in the 1970s in Professor Claude Allègre's laboratory at the Paris Institut de Physique du Globe, at a time when the discipline was really taking off in France. We also helped set up degree courses in geochemistry at the recently founded University of Paris 7, where we were appointed Assistant Lecturers. Our work together resulted in the publication of a short book in 1976, primarily for students, which quickly sold out and curiously enough was never reprinted! Few universities in those days offered specialist courses in geochemistry.

Times have clearly changed since then! Geochemistry is now taught in most universities and it is needless to recall here the fundamental contribution that this discipline has made to all areas of Earth sciences and cosmochemistry. It is always helpful, though, for students and for non-specialist faculty to have a textbook that provides a review of the basic concepts and the most recent contributions to the discipline. And this is the second reason why I am happy to present this book; because Francis Albarède's work fulfills both these requirements. The basic principles of the use of the chemical elements and their isotopes are set out clearly, together with their major applications in such varied domains as cosmochemistry (the formation of the chemical elements, of the Solar System, and of the planets), the internal dynamics of the Earth (with its various reservoirs and interaction among them, convection within the mantle, etc.), and its surface processes (hydrosphere, atmosphere, and climate change). Francis Albarède is particularly well qualified to deal with the diversity of geochemical applications because his own research has covered most of these major fields. A number of aspects that are sometimes overlooked in geochemistry books feature here, such as the processes of transport of elements (Chapter 4) or the concepts of residence time (in Chapter 5 on geochemical systems), and there is an overview of analytical techniques, which have proved so fundamental to the development of geochemistry. Moreover, the presentation is often novel (e.g. the presentation of geochronology, which is not just a catalog of the different methods), and the text is copiously illustrated with instructive diagrams and graphs.

One last reason why I particularly like this book is the method that Francis Albarède has chosen for setting out the principles underlying the main geochemical models: most of the relations describing these principles are demonstrated here, and the argument is invariably accompanied and supported by mathematical equations that unquestionably help the reader follow the reasoning. Advancing from one equation to the next is not always effortless, but the (slight) exertion required is well worth the trouble. Students need to discover or rediscover the satisfaction to be gained from working out the equations describing a particular process from what are often intuitive relationships and, above all, from understanding that such equations are a short-hand representation of an underlying physico-chemical model that it is often easy to symbolize through a simple diagram: in short, they need to call on their faculties of understanding and their aptitude for model-making rather than their memory. Reading this geochemistry textbook should encourage them to do just that. Readers may rest assured though, there is no need to know any advanced mathematics to understand this book; it is within the reach of any good college student. Nor is it devoid of humor: the allegory of dogs, and black and white cats to explain chemical fractionation and the absence of isotopic fractionation is a prime example! Through its resolutely model-based approach to processes and its concise and up-to-date explanations of the main contributions of the discipline, this book should become a standard text for college courses and a very valuable source of information for non-specialists eager to learn more about geochemistry. It comes out at a particularly fortunate time, just as new analytical instruments are about to widen the scope of geochemical tools substantially and give geochemistry a renewed impetus. I wish Francis Albarède's book swift and sustained success.

Professor Michel Condomines
Université de Montpellier II

Acknowledgments

I would like to thank Philippe Bonté, Dominique Boust, Hervé Cardon, Bill McDonough, Mireille Polvé, Yannick Ricard, Simon Sheppard, and Pierre Thomas for their advice. Careful reading of the original French manuscript by Janne Blichert-Toft, Fréderic Chambat, Michel Condomines, Don Francis, John Ludden, and Philippe Vidal, and by graduate students at ENS Lyon, has weeded out many errors of form and substance. Dave Manthey allowed me to reproduce graphics from his great Orbital Viewer. Agnès Ganivet kindly and effectively tidied up a first manuscript littered with syntactic errors. I would like to thank Nick Arndt, Edouard Bard, Janne Blichert-Toft, Marc Chaussidon, Al Hofmann, Dan Mckenzie, Bruce Nelson, Simon Sheppard, and Jacques Treiner, for reviewing the English manuscript. Chris Sutcliffe did an immense and wonderful job with translation and editing. Lesley Thomas was a clear-headed and efficient copy-editor.

In writing this book I have also sought to express my gratitude to the Institut Universitaire de France: since my appointment has allowed me to devote more time to research, I felt it only right that this should be requited by some concrete contribution to the teaching of geochemistry.

Introduction

Geochemistry utilizes the principles of chemistry to explain the mechanisms regulating the workings – past and present – of the major geological systems such as the Earth's mantle, its crust, its oceans, and its atmosphere. Geochemistry only really came of age as a science in the 1950s, when it was able to provide geologists with the means to analyze chemical elements or to determine the abundances of isotopes, and more significantly still when geologists, chemists, and physicists managed to bridge the chasms of mutual ignorance that had separated their various fields of inquiry. Geochemistry has been at the forefront of advances in a number of widely differing domains. It has made important contributions to our understanding of many terrestrial and planetary processes, such as mantle convection, the formation of planets, the origin of granite and basalt, sedimentation, changes in the Earth's oceans and climates, the origin of mineral deposits, to mention only a few important issues. And the way geochemists are perceived has also changed substantially over recent decades, from laboratory workers in their white coats providing age measurements for geologists or assays for mining engineers to today's perception of them as scientists in their own right developing their own areas of investigation, testing their own models, and making daily use of the most demanding concepts of chemistry and physics. Moreover, because geochemists generate much of their raw data in the form of chemical or isotopic analyses of rocks and fluids, the development of analytical techniques has become particularly significant within this discipline.

To give the reader some idea of the complexity of the geochemist's work and also of the methods employed, we shall begin by following three common chemical elements – sodium (Na), magnesium (Mg), and iron (Fe) – on their journey around system Earth. These three elements were created long before our Solar System formed some 4.5 billion years ago, in the cores of now extinct stars. There, the heat generated by the gravitational collapse of enormous masses of elementary particles overcame the repulsive forces between protons and triggered thermonuclear fusion. These reactions allowed particles to combine forming ever larger atomic nuclei of helium, carbon, oxygen, sodium, magnesium, and iron.

This activity is still going on before our very eyes as the Sun heats and lights us with energy released by hydrogen fusion. When, after several billion years, the thermonuclear fuel runs out, the smaller stars simply cool: the larger ones, though, collapse under their own weight and explode in one of the stellar firework displays that Nature occasionally stages, as with the appearance of a supernova in the Crab nebula in AD 1054. The matter scattered by such explosions drifts for a while in interstellar space as dust clouds similar to the one that can be observed in the nebula of Orion. Turbulence in the cloud and collisions between the particles makes the system unstable and the particles coalesce to form small rocky bodies known as planetesimals. Chondritic meteorites give us a pretty good idea of what these are like. Gravitational chaos amplifies very rapidly and the planetesimals collide to form larger and larger bodies with a star at the center of the arrangement, orbited by planets: a Solar System is born.

As the planet forms and evolves, our three chemical elements meet different fates. Sodium is a volatile element with a relatively low boiling point (881 °C) and large amounts of it are therefore driven off into space by the heat generated as the planet condenses. Iron, which is initially scattered within the rock mass, melts and collects at the heart of the planet to form the core, which, on Earth, generates the magnetic field. Magnesium has a boiling point of 1105 °C and behaves in an un-extraordinary way assembling with the mass of silicate material to form the mantle, the main bulk of the terrestrial planets, and residing in minerals such as olivine (Mg_2SiO_4), the pyroxenes ($Mg_2Si_2O_6$ and $CaMgSi_2O_6$), and, if sufficient pressure builds up, garnet ($Mg_3Al_2Si_3O_{12}$).

The chemical histories of the planets are greatly influenced by their magmatic activity, a term that refers to all rock melting processes. On our satellite, the Moon, the formation of abundant magmas in the first tens of millions of years of its history has left its mark in the surface relief. The crust, composed of the mineral plagioclase ($CaAl_2Si_2O_8$), is so light that it floats on the magma from which it crystallized, forming the lunar highlands that rise above the maria, the Moon's marias. The sodium that did not vaporize migrated with the magma toward the lunar surface and combined readily in the plagioclase alongside calcium. On Earth, water is abundant because temperatures are moderate and because the planet is massive enough for gravity to have retained it. The action of water and dissolved carbonates is another significant factor in the redistribution of chemical elements. Water causes erosion and so is instrumental in the formation of soils and sedimentary rocks. The presence of chemically bound water in the continental crust promotes metamorphic change and, by melting of metamorphosed sedimentary rocks, the formation of granite that is so characteristic of the Earth's continental crust. A substantial fraction of the sodium that did manage to enter into the formation of the early crust was soon dissolved and transported to the sea, where it has resided for hundreds of millions of years. Some marine sodium entered sediment and then, in the course of magmatic processes, entered granite and therefore the continental crust.

Magnesium, by contrast, tends to remain in the dense refractory minerals. It lingers in solid residues left after melting or precipitated during crystallization of basalt at mid-oceanic ridges or at ocean island volcanoes. Where it does enter fluids, it subsequently combines with olivine and pyroxene, which precipitate out as magma cools. Magnesium is stant to melting and is predominant in the mantle, which has ten to thirty times the magnesium content of the crust.

The Earth is a complex body whose dynamics are controlled by mechanisms that commonly work in opposing directions: differentiation mechanisms, on the one hand, maintained by fractionation of elements and isotopes between the phases arising during changes of state (melting, crystallization, evaporation, and condensation), and, on the other hand, mixing mechanisms in hybrid environments such as the ocean and detrital rocks that tend to homogenize components derived from the various geological units (rain water, granite, basalt, limestone, soil, etc.). By fractionation we mean that two elements (or two isotopes) are distributed in unequal proportions among the minerals and other chemical phases present in the same environment.

It can be seen then that the elements must be studied in terms of their properties in the context of the mineral phases and fluids accommodating them and in the context of the processes that govern changes in these phases (magmatism, erosion, and sedimentation). An understanding of transport mechanisms is very important in geochemistry. The term "cycle" is sometimes employed but we will see later that its meaning needs to be clarified. Conversely, the terminology used above, which emphasizes that sodium resides for a long time in the ocean or that magnesium lingers in solid residues of melting, illustrates that transfers among the different parts of the globe, such as the core, mantle, crust, and ocean, are to be considered kinetically, or, more loosely, dynamically, in terms of flows or transport rates. Radioactive decay, which alters the atomic nucleus and therefore the nature of certain isotopes at rates that are unaffected by the physical, crystallographic, or chemical environment in which they occur, will allow us to include the incessant ticking of these "clocks" in our study of these processes.

This book then moves on to the essential study of the dynamics and evolution of the mantle and continental crust, and the study of marine geochemistry and its implications for paleoclimatology and paleoceanography. A last chapter, specially written for the English edition, deals with the geochemical properties of a number of elements. Most students deplore the lack of such a systematic approach in the literature: what is supposed to be common knowledge is never taught, simply because the odds of being inaccurate, unbalanced, and superficial are too great. Appendix F provides an overview of a number of methods for analyzing chemical elements and isotopes.

It will probably be found that this book relies more heavily on equations than most other geochemistry textbooks. I maintain that a proper scientific approach to our planet must use all the available tools, especially those of physics and chemistry, to supplement purely descriptive and analytical approaches. I therefore ask readers to persevere despite the superficial difficulty of some of the equations, which are, after all, nothing more than a means of encoding concepts that ordinary language is powerless to convey with adequate precision. In the process of translating the French textbook into English, I realized that about 15% of the words had been irretrievably lost due to the verbose nature of the French language so, to keep my publisher happy, I have added a substantial amount of text and a number of figures. Still, there are two absences of note: organic geochemistry and atmospheric geochemistry. I can offer no excuse other than that my own lack of familiarity with these areas would make me an unsuitable guide.

Oscar Wilde said that "... nothing that is worth knowing can be taught." A collection of observations is no more science than a dictionary is literature. Above all, I have tried in this book not to bury the reader beneath a mountain of facts and theories. Challenges to today's "facts" will fuel tomorrow's theories. Scientific method requires that the formulation of incisive questions on the basis of observations must alternate with the collection of new data. These data will in turn confirm or falsify the predictions of the quantitative models.

Many scientists consider writing textbooks a menial chore. Moreover, the craze for photocopying means that an author's lot is not a happy one financially. No more than gain is fame the spur either: my previous book, published by Cambridge and which, I am told, sold well, ranks in 322 368th place on the amazon.com best-sellers list! I hope that academic staff will have a road map for introductory college courses in geochemistry. I have been asked so many times by genuinely motivated colleagues from other disciplines where a compact description of geochemistry could be found. I also hope this short textbook will meet this specific demand. But the devil is in the detail and I occasionally had to cut corners: some concepts, definitions, and proofs could have benefited from more rigor and from more detailed supporting arguments. I very much wanted to keep this book short, I hope that my specialist colleagues will forgive the short-cuts used to this effect.

Exercises accompanying this book will be posted on a website that will be advertised on the author's personal page: http://www.ens-lyon.fr/~albarede. Readers may e-mail any queries, criticisms and – who knows – words of encouragement to the author (albarede@ens-lyon.fr).

Chapter 1
The properties of elements

The 92 naturally occurring chemical elements (90, in fact, because promethium and technetium are no longer found in their natural state on Earth) are composed of a nucleus of subatomic nucleons orbited by negatively charged electrons. Nucleons are positively charged protons and neutral neutrons. As an atom contains equal numbers of protons and electrons with equal but opposite charges, it carries no net electrical charge. The mass of a proton is 1836 times that of an electron. The chemical properties of elements are largely, although not entirely, determined by the way their outermost shells of electrons interact with other elements. Ions are formed when atoms capture surplus electrons to give negatively charged anions or when they shed electrons to give positively charged cations. An atom may form several types of ions. Iron, for example, forms both ferric (Fe^{3+}) ions and ferrous (Fe^{2+}) ions, while it also occurs in the Fe^0 elemental form.

A nuclide is an atomic nucleus characterized by the number Z of its protons and the number N of its neutrons regardless of its cloud of electrons. The mass number A is the sum of the nucleons $N + Z$. Different interactions act in the nucleus and explain its binding: the short-range (nuclear) strong force, the electromagnetic force, and the mysterious weak force. Two nuclides with the same number Z of protons but different numbers N of neutrons will be accompanied by the same suite of electrons and so have very similar chemical properties; they will be isotopes of the same element. The "chart of the nuclides" (Fig. 1.1) shows that in order to be stable, nuclides must contain a specific proportion of neutrons and protons. For light elements ($Z > 40$), the number N of neutrons and the number Z of protons in a nucleus tend to be

Figure 1.1: Chart of the
nuclides (overview). Stable
elements have approximately
the same number Z of
protons and number N of
neutrons: this relationship
defines the valley of stability.
Elements that deviate
significantly from this rule are
unstable (radioactive).

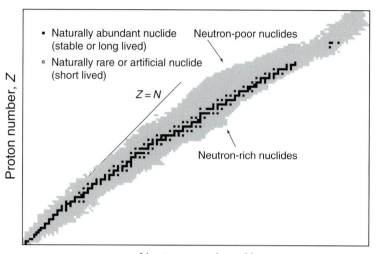

Figure 1.1: Chart of the nuclides (overview). Stable elements have approximately the same number Z of protons and number N of neutrons: this relationship defines the valley of stability. Elements that deviate significantly from this rule are unstable (radioactive).

similar: since attraction between proton and neutron is slightly stronger than proton–proton and neutron–neutron attraction, a minimum energy is reached for $N = Z$. At higher masses, the Coulomb repulsion between protons makes it energetically favorable to have $N > Z$: a good approximation of the "valley of stability" is $N = Z + Z^2/160$.

Nuclei with N and Z too far from this valley of stability are unstable and are said to be radioactive. An isotope is radioactive if its nucleus undergoes spontaneous change such as occurs, for instance, when alpha particles (two protons and two neutrons) or electrons are emitted. It changes into a different isotope referred to as radiogenic by giving out energy, usually in the form of gamma radiation, some of which is hazardous for humans. Several Internet sites provide tables of all stable and radioactive nuclides, e.g. http://www2.bnl.gov/ton. The vast majority of natural isotopes of naturally occurring elements are stable, i.e. the number of their protons and neutrons remains unchanged, simply because most radioactive isotopes have vanished over the course of geological time. They are therefore not a danger to people.

1.1 The periodic table

The atomic number of an element is equal to the number of its protons. We have seen before that the atom's mass number is equal to the number of particles making up its nucleus. The Avogadro number $\mathcal{N}$ is the number of atoms contained in 12 g of the carbon-12 isotope. The atomic mass of an isotope is the weight of a number $\mathcal{N}$ of atoms of that isotope. D. I. Mendeleev's great discovery in 1871 was to demonstrate the

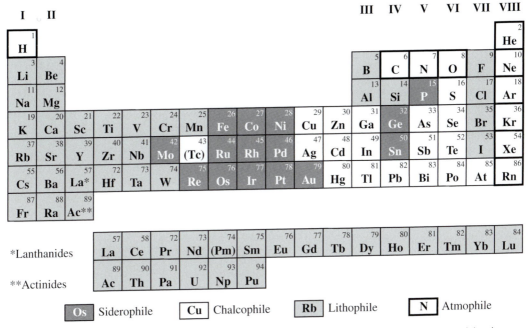

Figure 1.2: Mendeleev's periodic table of the elements and their geochemical classification after Goldschmidt. The elements in parentheses do not occur naturally on Earth. The atomic number of each element is given. Roman numerals over columns indicate groups.

periodic character of the properties of elements when ordered by ascending atomic number (Fig. 1.2). Melting point, energy of formation, atomic radius, and first ionization energy all vary periodically as we work through Mendeleev's table. The geochemical properties of elements are reflected by their position in this table. The alkali metals (Li, Na, K, Rb, Cs), alkaline-earth metals (Be, Mg, Ca, Sr, Ba), titanium group elements (Ti, Zr, Hf), but also the halogens (F, Cl, Br, I), inert gases (He, Ne, Ar, Kr, Xe), rare-earths (lanthanides), or actinides (uranium family) all form groups sharing similar chemical properties; indeed these properties are sometimes so similar that it was long a challenge to isolate chemically pure forms of some elements such as hafnium (Hf), which was only separated from zirconium (Zr) and identified in 1922.

It is therefore very important to understand how elements are ordered in the periodic table. Put simply, an atom can be represented as a point-like nucleus containing the mass and charge of the nuclear particles and by electrons orbiting this point. Quantum mechanics requires the different forms of electron energy to be distributed discretely, i.e. at separate energy levels. This discrete arrangement applies both to the total energy and to the angular momentum of the electrons on their atomic orbitals. An orbital corresponds to a probability distribution of the electron around the nucleus. An orbital (Fig. 1.3) is denoted by a set of integers known as quantum numbers. The first quantum number n

The properties of elements

Figure 1.3: Examples of orbital geometry. Shown here are the surfaces of maximum probability of electron localization around the nucleus corresponding to various orbitals. s, p, d, f are the quantum numbers. Note the two types of d orbitals. Drawn using Orbital Viewer (Dave Manthey).

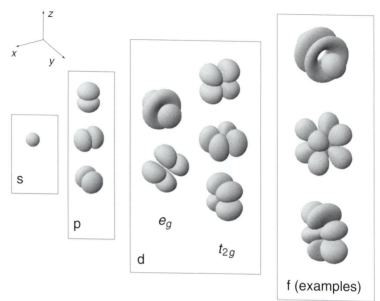

defines the general energy level of the electron and can take positive values 1, 2, 3, etc. It defines the main electron shells, which are sometimes represented by the letters K, L, M, etc. The second quantum number l ranges from 0 to $n - 1$ and characterizes the shape of the orbitals of each shell, which are usually designated by the letters s, p, d, f. The third number m $(0, \pm 1, \ldots, \pm l)$ describes this shape in the presence of a magnetic field, and the fourth number s describes the relative direction of spin of the electron around its own axis and around the orbit. The Pauli exclusion principle states that no two electrons can have the same quantum number. Mendeleev's table can be constructed by assigning a unique set of quantum numbers to each element (Table 1.1).

A number of Internet sites provide detailed periodic classifications, of which I can recommend http://www.shef.ac.uk/chemistry/web-elements, while Dave Manthey's excellent site http://www.orbitals.com/orb/ov.htm provides software to create very professionally drawn orbital pictures (Fig. 1.3).

In the periodic table, columns I (alkali metals) and II (alkaline-earth metals) correspond to the filling of s orbitals, and columns III to VIII to that of the p orbitals. The intermediate columns (transition elements such as iron and platinum) differ in the occupation of their d orbitals. When occupied, these d orbitals are normally closer to the nucleus than the s orbitals of the next shell out. Occupation of the orbitals is noted nx^i, where x represents the type of orbital (s, p, d, f), n its principal quantum number and i the number of electrons it contains. Most elements of the

Table 1.1: *Electronic configuration of the light elements*

Element	Col.	Quantum numbers				Configuration
		n	l	m	s	
H	I	1	0	0	+1/2	$1s^1$
He	VIII	1	0	0	−1/2	$1s^2 = $ [He]
Li	I	2	0	0	+1/2	[He] $2s^1$
Be	II	2	0	0	−1/2	[He] $2s^2$
B	III	2	1	−1	+1/2	[He] $2s^2\,2p^1$
C	IV	2	1	−1	−1/2	[He] $2s^2\,2p^2$
N	V	2	1	0	+1/2	[He] $2s^2\,2p^3$
O	VI	2	1	0	−1/2	[He] $2s^2\,2p^4$
F	VII	2	1	+1	+1/2	[He] $2s^2\,2p^5$
Ne	VIII	2	1	+1	−1/2	[He] $2s^2\,2p^6 = $ [Ne]
Na	I	3	0	0	+1/2	[Ne] $3s^1$

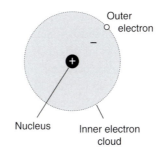

Figure 1.4: Shielding of the nuclear charge by the cloud of electrons orbiting between the outer electrons and the nucleus.

first series (e.g. V, Cr, Mn, Fe, Co, Ni, Cu, Zn) have an electron formula of the type [Ne]$3s^2\,3p^6\,3d^n\,4s^2$, where [Ne] represents the fully occupied orbitals of a neon atom, and their divalent ions, such as Fe^{2+} and Cu^{2+}, have a configuration [Ne]$3s^2\,3p^6\,3d^n$. These transition elements differ only by the number n of electrons in orbital 3d but have an identical outer electron shell 4s, which explains why their chemical properties are so similar. This phenomenon is further amplified in the rare-earths (or lanthanides), such as La and Ce (shell 4f), and the actinides (5f), like U and Th, where the s and p orbitals of the external shells are identical. The electrostatic attraction of the outer electrons by the nucleus is reduced by the cloud of the inner electrons (Fig. 1.4), a phenomenon known as shielding. For the lanthanides and the actinides, the f electrons on their multi-lobate orbitals (Fig. 1.3) do not screen the increasing charge of the nucleus as efficiently as the more regularly shaped lower-order orbitals. As a result, their atomic radii decrease smoothly with their increasing atomic number, a phenomenon known as lanthanide (and actinide) contraction.

1.2 Chemical bonding

Atoms and ions combine to form matter in its solid, liquid, or gaseous states. The importance of occupation of the outer electron shell can be illustrated by comparing the interaction between two atoms of helium, where two electrons occupy orbital 1s, with the interaction between two hydrogen atoms, each with a single electron only. When the two helium atoms come close together and their electron clouds interpenetrate, one atom's electrons cannot be accommodated by the orbital of the other as this would infringe the Pauli exclusion principle. They must therefore jump to orbital 2s, at a cost in energy that penalizes the formation of such bonds. Two hydrogen atoms, however, can lend one another an electron and orbital 1s. The resulting electron configuration is more advantageous than that of isolated hydrogen atoms and chemical bonding is favored. The type of chemical bonding is determined by the extent to which electrons are shared and by the time they spend away from their atom of origin. If the electron is transferred permanently, ionic bonding has occurred: a sodium atom in the presence of a chlorine atom will give up its isolated 3s electron because their outer shells will then be completely occupied, with the sodium configured like neon and the chlorine like argon. The ions formed in this way, noted as Na^+ and Cl^-, are particularly stable; their outer electron shell is largely spherical and these ions act like electrically charged spheres mutually attracted by their electrostatic fields to form ionic compounds such as table salt, NaCl. Conversely, when the number of electrons that can be exchanged fails to fill the outer shells of the two partners, a covalent bond is formed. Two hydrogen atoms lend one another their missing electron but must share the two 1s electrons over time by forming hybrid orbitals of complex geometry thereby allowing them to fill the outer shell of both atoms.

1.3 States of matter and the atomic environment of elements

Bonds formed by condensed materials are generally more complex than those formed by gases. In the silicates, which are so important to our understanding of geological phenomena, a small silicon (or possibly

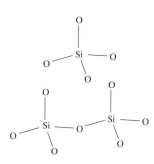

Figure 1.5: SiO_4 tetrahedra, which are the building blocks of silicate structures and their polymerization. Aluminum similarly forms AlO_4 tetrahedra.

aluminum) atom (Fig. 1.5) lies at the center of a tetrahedron of four oxygen atoms. As in carbon chemistry, SiO_4 tetrahedra may polymerize to varying degrees by sharing one or more oxygens at their apexes. The Si–O bond is a quite strongly covalent one. Other elements, such as Mg, Fe, or Na, may be accommodated within the silicate framework in their ionic forms Mg^{2+}, Fe^{2+}, Na^+.

The many crystallized silicate and alumino-silicate structures are classified according to the pattern formed by their tetrahedra. The most important ones in geology are:

1. Isolated-tetrahedra silicates: the most common minerals in this family are the olivines, such as forsterite Mg_2SiO_4, and garnets, such as pyrope $Mg_3Al_2(SiO_4)_3$.
2. Single-chain silicates: these are the pyroxenes, which fall into two groups with two different crystallographic systems; orthopyroxenes, such as enstatite $Mg_2Si_2O_6$, and clinopyroxenes, such as diopside $CaMgSi_2O_6$.
3. Double-chain silicates: amphiboles, such as tremolite $Ca_2Mg_5Si_8O_{22}(OH)_2$ or hornblende $Ca_2Mg_4Al_2Si_7O_{22}(OH)_2$. The formation of these hydroxy-lated minerals requires some degree of water pressure.
4. Sheet silicates: micas and clay minerals usually containing aluminum, potassium, and smaller ions such as Fe^{2+} and Mg^{2+}. A distinction is drawn between di-octahedral micas like muscovite (common white mica) $K_2Al_6Si_6O_{20}(OH)_4$ and tri-octahedral micas like biotite (ordinary black mica) $K_2Mg_6Al_2Si_6O_{20}(OH)_4$, the difference being in the number of sheets in their basic pattern. This family is extremely diverse.
5. Framework silicates: these silicates are interconnected at each of their apexes. The family includes quartz SiO_2 and the feldspars, the most important of which are albite $NaAlSi_3O_8$, anorthite $CaAl_2Si_2O_8$, and the various potassium feldspars whose formula is $KAlSi_3O_8$.

Other significant minerals include iron oxides and titanium oxides, which are commonly cubic ionic solids such as magnetite Fe_3O_4 and ilmenite $FeTiO_3$. Corundum is the oxide of aluminum Al_2O_3. Calcium carbonate (calcite, aragonite) and magnesium carbonate (magnesite), formed by stacking of rhombohedral Ca^{2+} or Mg^{2+} ions, contain small carbonate groupings CO_3^{2-} able to rotate around the axis of symmetry.

The minerals cited above are only examples, albeit important ones, but alone they fail to provide a sufficiently precise representation of the chemical diversity of rocks. Elements are located within minerals at sites characterized by the number of oxygen atoms that they have as their immediate neighbors. We have already come across the tetravalent silicon ion Si^{4+} (i.e. carrying four positive charges) at a tetrahedral site and said that it could be replaced by the trivalent aluminum ion Al^{3+}. Electrical neutrality is maintained through paired substitutions such as $Al^{3+}Al^{3+}$ substituting for $Si^{4+}Mg^{2+}$. Such substitution is possible and even commonplace in pyroxenes, amphiboles, micas, and feldspars as the two ions have similar ionic radii (0.39 and 0.26 Å, respectively) and similar electrical charges. Likewise, Mg^{2+} ions (0.72 Å) located at octahedral sites of naturally occurring olivine, i.e. those surrounded by six oxygen atoms, are commonly replaced by Fe^{2+} ions (0.61 Å) or Ni^{2+} ions (0.69 Å). Ions of the rare-earth element family, such as the Yb^{3+} ytterbium ion (0.99 Å), may substitute for the Ca^{2+} ion (1.00 Å) in clinopyroxenes or amphiboles. As these continuous substitution phenomena are analogous to those whereby various ions co-exist in aqueous solutions, the term solid solution is used.

Not all solid solutions are possible. The larger ions, such as the alkali metals (K^+, Rb^+) or the alkaline-earth metals (Sr^{2+}, Ba^{2+}) of the higher periods, have ionic radii that are too big for them to fit readily into the common silicate minerals. Ions with different charges manage to substitute for major ions if the electrical imbalance can be offset locally: a $(Ca^{2+})_2$ pair may thus be replaced by a $Yb^{3+}Na^+$ pair in a clinopyroxene, whereas its replacement by a Th^{4+} thorium ion requires the formation of a defect with a high energy cost. Ions that carry too high a charge or whose ionic radius is too small or too large are rejected by the lattice of essential minerals and concentrate either in accessory minerals, such as the phosphates or titanates, or in poorly characterized phases in grain fractures and interstices. These outcasts are termed incompatible elements. However, this is a relative concept: potassium and barium are incompatible in the feldspar-free mantle, yet, in the continental crust, where feldspar is abundant, K and Ba are compatible. The volatile elements and compounds, such as the rare gases, water, and CO_2, call for some attention. As long as no gas phase is present, i.e. as long as the concentration of one of the most abundant volatiles in the solid or liquid in question does not exceed saturation level, they behave like any other element with varying levels of compatibility. In the absence of vapor, for example, the inert gases like helium or argon need not be classified

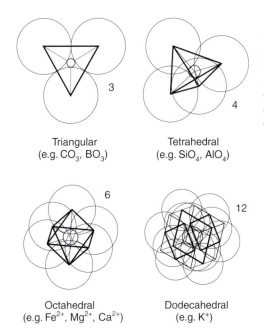

Figure 1.6: The main ion coordination systems in naturally occurring minerals: triangular (three closest neighbors), tetrahedral (4), octahedral (6), and dodecahedral (12) coordination.

Triangular
(e.g. CO_3, BO_3)

Tetrahedral
(e.g. SiO_4, AlO_4)

Octahedral
(e.g. Fe^{2+}, Mg^{2+}, Ca^{2+})

Dodecahedral
(e.g. K^+)

separately from the other trace elements.

The energy stored in silicates as chemical bonds depends on the nature of the cations and the crystal sites accommodating them. Depending on their ionic radius, cations may occupy sites of varying size and the number of oxygen neighboring atoms with which they bond (coordination number) increases with the size of the site. Carbon and boron atoms combine with oxygen in a triangular arrangement (i.e. threefold coordination, Fig. 1.6), while silicon and aluminum atoms combine with oxygen to form a tetrahedron (fourfold coordination). However, medium-sized cations, such as Fe^{2+}, Mg^{2+}, or Ca^{2+}, will take up vacant octahedral sites (sixfold coordination) between SiO_4 tetrahedra while the biggest ions, such as K^+ or OH^- hydroxyl anions, require the most spacious sites, normally of twelvefold coordination.

The transition elements (V, Fe, Cu, Zn, etc.) are particularly sensitive to their crystalline environment. They differ from each other by a different filling of their d orbitals (Fig. 1.3). Two of these orbitals, called e_g, have their lobes lying along the axes of rectangular coordinates, while three orbitals, called t_{2g}, lie in each plane defined by two axes but with their lobes sticking out in between the axes. In the most common octahedral sites occupied by transition elements in silicates, with an oxygen atom at each apex along the axes, an electron occupying a t_{2g} orbital feels the repulsion by the electrons from the oxygen ions,

Figure 1.7: Crystal field effect on a transition element, such as Fe, Mn, Cr, in octahedral coordination. Six oxygen atoms form the apexes of the octahedron. The t_{2g} orbitals of the transition element lie between the oxygen atoms, while the e_g orbitals point toward them. The filling of an e_g orbital by electrons therefore has to overcome excess repulsion energy compared with a t_{2g} orbital.

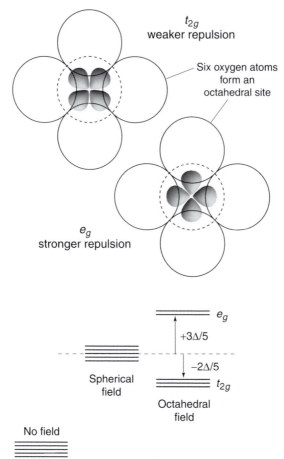

t_{2g}
weaker repulsion

Six oxygen atoms form an octahedral site

e_g
stronger repulsion

e_g

$+3\Delta/5$

$-2\Delta/5$

Spherical field

t_{2g}

Octahedral field

No field

Figure 1.8: Splitting of the d energy levels in the octahedral site of Fig. 1.7. If the ion is inserted into a spherical site, the repulsion to overcome is symmetric for the electrons on all the d orbitals. In an octahedral site, the two e_g orbitals are subjected to stronger repulsion than the three t_{2g} orbitals. For the same overall interaction energy, and a difference Δ in bonding energy between t_{2g} and e_g, the energy shift of the t_{2g} orbitals is $-2\Delta/5$, while the shift for the e_g orbitals is $+3\Delta/5$. In a tetrahedral site, the situation would be reversed.

the so-called crystal field, much less than an electron occupying an e_g orbital (Fig. 1.7). Calling Δ the difference in bonding energy between t_{2g} and e_g, the energy shift of the three t_{2g} orbitals is $-2\Delta/5$, while the shift for the two e_g orbitals is $+3\Delta/5$, thereby ensuring that the mean energy shift with respect to a spherical environment is zero (Fig. 1.8). Let us consider the trivalent chromium ion Cr^{3+} with electronic formula $[Ar]3d^3\ 4s^0$ ([Ar] stands for the orbital filling of argon). One electron on each t_{2g} orbital gives this ion a bonding energy of $3 \times (-2\Delta/5) = -6\Delta/5$. For divalent nickel Ni^{2+} ($[Ar]3d^8\ 4s^0$), the first three electrons occupy the three t_{2g} orbitals, the next two electrons fill the two e_g orbitals, then the last three complete the filling of the t_{2g}. For Ni^{2+}, the energy gain is $6 \times (-2\Delta/5) + 2 \times (3\Delta/5) = -6\Delta/5$. Cr^{3+} and Ni^{2+} therefore snuggle in octahedral sites and are notably enriched in silicates such as olivine and pyroxenes. Other elements, such as Mn^{2+} ($[Ar]3d^5 4s^0$) and Zn^{2+} ($[Ar]3d^{10}4s^0$), have no energy gain and will show no preference between sites. The converse situation is true for these ions in tetrahedral coordination.

Pressure is especially important in the environment of ions. Oxygen is more compressible than smaller ions and, with its cation coordination number increasing, it plays an essential role. Thus, at depths of more than 660 km, pressure levels in the lower mantle are such that silicon co-ordination shifts from fourfold to sixfold thereby promoting the compact stacking of oxygen atoms. This is a general process and it is therefore difficult to predict with any accuracy the chemical properties of elements at very great depths.

1.4 Geochemical classifications

There are many ways to arrange elements by their geochemical prop-erties. Such practice, reducing as it does the wide diversity of behavior of elements to limited ranges of behavior, might be thought pointless, but it does provide an overview of their chemical properties. The most widespread classification is probably that of Goldschmidt (Fig. 1.2), based primarily on the energy of formation of oxides and sulfides, and which is operationally useful. The lithophile elements (Na, K, Si, Al, Ti, Mg, Ca) generally concentrate in the rock-forming minerals of the crust and mantle; the siderophile elements (Fe, Co, Ni, Pt, Re, Os) have an affinity for iron and therefore concentrate in the Earth's core; the chal-cophile elements (Cu, Ag, Zn, Pb, S) readily form sulfides; the atmophile elements (O, N, H, and the inert gases) concentrate in the atmosphere. Certain elements in each group tend to be volatile; K is a more volatile lithophile than either Mg or Ti (see Chapter 9). Refractory elements such as Mg or Cr tend to concentrate in solid residues.

The charge and ionic radius of an ion define the electrical potential in the neighborhood of this ion and are often considered together. Large ions carrying weak charges (K, Rb, Cs, Ba) are accommodated with diffi-culty by the main minerals of the mantle, except for the K-feldspars, and so are concentrated in the continental crust. They are known as large-ion lithophile elements (LILE). Small ions carrying strong charges (Zr, Nb, Th, U) develop intense electrostatic fields (high field-strength elements, HFSE), hence they do not readily substitute for the major elements in or-dinary minerals. However, relative positions in the periodic table remain essential, and elements in the same column are geochemically similar, as is the case of the alkali metals (Na, K, Cs), alkaline-earth metals (Mg, Ca, Sr), halogens (F, Cl, Br), or transition elements of the same family.

1.5 The different reservoirs and their composition

In solids, elements are sequestered in minerals, minerals assemble as rocks, and rocks constitute the main geodynamic units of the mantle and the crust. Some elements occur in particularly large abundances in seawater or in the atmosphere (H, O, N, Ar). Terrestrial rocks fall into three main categories: igneous rocks, such as basalts and granites, produced by magmatic processes, i.e. from the melting of rocks; sedimentary rocks, formed by the accumulation of clastic and biological particles or by chemical precipitation on the floors of the oceans and other bodies of water; metamorphic rocks, produced by "firing" existing rocks at high temperature, under pressure, and, for most of them, in the presence of aqueous or carbonic fluids. Extra-terrestrial rocks cannot be classified in quite the same way. Igneous rocks occur on the Moon, most probably on Mars, and as meteorites (achondrites). Another type of meteorite, known as a chondrite, has no terrestrial equivalent. Chondrites were formed by condensation of gases from the solar nebula and by droplets of silicate liquids called chondrules, from which their name is derived. Metamorphic transformations may affect planetary rocks and meteorites as well.

A reservoir is a loosely defined term referring to a very large body of rock (mantle, crust), water (ocean), or gas (atmosphere) whose mean composition stands in sharp contrast to the composition of other reservoirs. It may contain a variety of components, but their composition is normally very different from the composition of components present in other reservoirs: for instance, the Si-rich rocks making up the continental crust reservoir are easily differentiated from the Mg-rich rocks making up the mantle. A reservoir may be spatially continuous (e.g. the ocean) or scattered over large distances in the Earth (e.g. recycled oceanic crust in the deep mantle). The most abundant chemical elements in each of the main terrestrial reservoirs are listed in Appendix A. However, a clearer understanding of these chemical distributions requires some idea of which minerals contain these elements (see Chapter 10). Oxygen is found just about everywhere, whereas silicon is confined to the silicates, which are by far the most abundant minerals. The silicon content of minerals is a particularly significant parameter because the SiO_2 (silica) concentration is a measure of the "acidity" of rocks: this obsolete term, which dates from the days when silicates were viewed as silicic acid salts, still pervades the literature. The silica concentration of minerals increases from olivine, pyroxene, amphibole, and mica through to feldspar and quartz. Magnesium and iron are particularly abundant in olivine, in pyroxene of igneous rocks, in amphibole, and in the sheet minerals (biotite, chlorite, and serpentine) of metamorphic rocks. A felsic (acidic) rock, such as a granite or a rhyolite, is rich in silicon and poor in magnesium and iron; a mafic rock, the archetype of which is basalt, has a high Mg and Fe content. Calcium is found, above all, in igneous pyroxene and calcitic feldspar (plagioclase), in sedimentary carbonate, and in metamorphic amphibole. Aluminum has many carriers: in igneous rocks it is located, by increasing order of pressure, in plagioclase, spinel (oxides), and garnet; it concentrates in the clay minerals of sedimentary rocks and in mica of metamorphic rocks (biotite and muscovite). Potassium and sodium are scarcely found outside mica and feldspar.

When rock melts, some elements (Na, K, Al, Ca, Si) are fusible, whereas others (Mg and to a lesser extent Fe) are more refractory; magmatic melting therefore contributes to geochemical fractionation among reservoirs. In the same way, some elements (Na, K, Ca, Mg) are more soluble in water than others, inducing further geochemical fractionation during erosion and sedimentation.

When the composition of the mantle is compared with that of the Earth as a whole it can be seen to have a high refractory-element content, especially of Mg and Cr, and a lower content of fusible elements, especially Na, K, Al, Ca, and Si, demonstrating its residual character with regard to melting. As might be expected, olivine and pyroxene are predominant in the mineralogy of the upper mantle: peridotite is the ubiquitous rock forming the upper mantle. The continental crust, on the other hand, is enriched in fusible elements (accommodated mainly in feldspar, quartz, and clay minerals) and exhibits a melt "liquid" character in contrast to the residual mantle. The oceans are obviously enriched in soluble Na, K, and Ca cations and anions (Cl, SO_4), while elements that are both insoluble and fusible (Si, Fe, and Al) accumulate in clastic sedimentary rocks (clays).

The composition of the Sun, the Earth's crust and mantle, etc., is given in Appendix A. It is not always easy to determine these compositions; while observation of the solar spectrum and analysis of meteorites, of seawater, and of river water yield data that can be tabulated fairly directly, determining the composition of the Earth's crust calls for discussion of the nature of the lower crust, the lower mantle, the core, and of the Earth as a whole. The mechanisms responsible for forming these major, but not directly observable, geological reservoirs must therefore be reasonably well understood before their compositions can be estimated. This will be covered briefly in Chapter 8.

1.6 The nucleus and radioactivity

Although the forces holding the nucleus together are extremely powerful, observation shows that some nuclei are unstable, i.e. radioactive.
Radioactivity is a property of the nucleus and involves energies in the
order of one MeV (1 megaelectron-volt $= 1.6 \times 10^{-13}$ J) per nucleon. It
is not dependent on the atom's electron suite and therefore on any chemical processes, which take place at much lower energies: removing the
first electron from an atom typically requires only 1–10 eV (ionization
potential) per atom. Mineralogical, temperature, and pressure processes
involve even smaller energies. Radioactivity is therefore independent of
the chemical and mineralogical environment of the element, of temperature, which is a measurement of its excitation, and of pressure, which affects only the electron shells. For example, the probability of rubidium-87
or uranium-238 decaying per unit of time is the same in the ocean or in
a lake, in a granite or a limestone, in the Earth's crust or lower mantle,
on the Moon or on Mars, etc. This probability has not varied measurably
over the course of geological time. The remarkable consistency of radiometric isotopic chronometers of diverse geologic and planetary objects
obtained by using different radioactive isotopes dispels any doubt about
the validity of nuclear clocks. The extreme temperature conditions found
in massive stars provide a few exceptions to this rule, but these are of no
practical relevance to our Solar System.

There are a number of decay processes:

1. The α (alpha) process, which is the emission of a helium nucleus (two protons
   and two neutrons), is common at high mass. The nucleons (positively charged
   protons and neutral neutrons) are held together by the short-range attractive
   strong force. This force is essentially restricted to adjacent nucleons and
   therefore varies linearly with their number. In contrast, the electromagnetic
   force, which tends to pry the protons apart, is weaker but acts over a broader
   range, so that it involves the entire nucleus: it varies with the total number
   of proton pairs and therefore with the squared number of protons. Overall,
   for the heavier nuclei, the repulsive force therefore tends to compensate the
   attractive force and the mean energy holding the particles together decreases.
   The α (alpha) process occurs when the sum of the masses of the daughter
   nuclide and the α particle is less than the mass of the parent nuclide. It is a
   consequence of a quantum property of nuclear particles known as the tunnel
   effect: although the potential barrier opposing ejecting an α particle is much
   greater than the energy gain resulting from separation, there is a non-zero
   probability that an ejection will occur. For example $^{147}Sm \rightarrow {}^{143}Nd + \alpha$.
2. The β^- (beta minus) process involves the emission of an electron by the
   parent nucleus. As indicated above, the lower energy of the proton–neutron
   interaction with respect to that of similar nucleons favors a nucleus with
   an equal number of protons and neutrons. When $N > Z$, this condition

is violated, and excess energy is released by converting a neutron into a proton and an electron. This process calls for a special type of force, the weak interaction, which is different in nature from both gravitational and electromagnetic forces. It is of little relevance to our familiar environment and is difficult to explain simply. The continuous distribution of electron energy emitted during this process cast doubt on the quantification of nuclear energy until it was shown that there was a particle of zero mass, the neutrino, that practically does not interact with the matter through which it passes, but which can carry considerable energy. Collapsing stars lose energy through neutrino emission. For example, $^{87}Rb \rightarrow {}^{87}Sr + \beta^- + \nu$, where β^- is the electron and ν the neutrino. β^- radioactivity is a common process when the nucleus has a high neutron/proton ratio. A symmetrical process of β^+ (beta plus) positron emission (particle with the same mass as an electron but of opposite charge) takes place when the nucleus has a high proton/neutron ratio ($N < Z$), an unusual situation for natural nuclides.

3. Capture of an electron of the K shell is a less frequent process, and was identified by von Weizsäcker during his investigation of excess argon-40 in the Earth's atmosphere. For example, $^{40}K + e^- \rightarrow {}^{40}Ar$. Electron capture affects the nuclide in much the same way as positron emission does.

4. Spontaneous fission of some heavy atoms like uranium-238 or plutonium-244 is a rather rare and very slow process; it forms the basis of the fission-track dating method.

A nuclide may be unstable with respect to two decay processes and the probabilities of decomposition by each process are additive. For example, potassium-40 (^{40}K) decays dually into ^{40}Ca by β^- emission and into ^{40}Ar by electron capture; this is known as a branched mode of radioactive decay.

Chapter 2

Mass conservation – elemental and isotopic fractionation

Before discussing the composition of the different systems of geological interest and the exchanges of matter that make those systems evolve relative to each other, it is worth recalling the principles governing the geochemical differentiation of our planet. These seemingly simple principles conceal what are often daunting complexities. They are: the principle of conservation of mass, elementary and isotopic fractionation induced by phase changes, kinetic fractionation, and radioactivity.

In the Introduction, we alluded to the contrast between mixing processes and differentiation processes. We will now look at a number of examples.

1. Partial fusion of the mantle beneath mid-ocean ridges produces basaltic liquids whose chemical composition is different from the ultramafic chemical composition of the source peridotite. This chemical fractionation of elements between the molten fluid and its parent medium can be described by thermodynamic rules. The former makes up the oceanic crust while the latter forms the refractory base of the lithosphere (located in the oceanic plates beneath the crust). When the oceanic crust and oceanic lithospheric mantle plunge at subduction zones, they begin a long journey within the mantle, where convection folds and stretches them, in much the same way as a baker kneads dough, progressively eliminating the differences that initially existed between the two constituent parts. Convective mixing therefore undoes the effect of magmatic differentiation.

2. The erosion of a granite yields clay minerals and quartz grains, and adds to the dissolved load of run-off water; the clay accumulates on the ocean floor, the quartz is deposited as sandstone on the continental shelf or slope, and

the river water mixes with the ocean waters. Through the process of erosion, a granite body of initially homogeneous composition is separated into three different products that meet very different fates. Over the course of time the sediments laid down are scraped off along subduction zones, where they are buried under considerable thicknesses of rock. There tectonic deformation and metamorphic mineralogical transformations at high temperature and high pressure, combined with fluid action, provide conditions amenable to the partial melting of crustal rocks (anatexis), and the siliceous fluids formed escape toward the surface producing granite intrusions. The work of differentiation performed by erosion is therefore destroyed by metamorphic and anatectic recombination.

3. In the ocean, chemical differentiation occurs at the surface, as biological activity draws on nutrient elements as well as calcium carbonate and silica. The re-dissolution of dead organisms and waste materials, together with bottom-water circulation, remixes the components that were separated out by biological activity. Once again the opposing processes of differentiation and mixing are at work.

We turn now to the methods of mass balance analysis, which are the very essence of these differentiation and mixing processes. By such methods we can determine the relative abundances of the constituent parts mixed together in a composite product and evaluate the proportions of phases that have separated out in a differentiation product.

First some essential definitions. Any system, whether natural or artificial, is made up of components and species. Components are chemical entities (atoms, ions, or, for a rock, metal oxides, such as Na, Na^+, Na_2O) from which the chemical composition of the rock can be described fully and uniquely. Rock compositions are commonly given in weight percent of its constitutive oxides SiO_2, Al_2O_3, Na_2O, etc., while the composition of solutions is normally given as moles of ionic component, Na^+, Ca^{2+}, Cl^-, etc., per kilogram of water. One unit system is as good as another, provided consistency is maintained. The set of components itself may not be unique: a rock can be described as proportions of atoms or as proportions of oxides. Components are not normally found as such in the system; a familiar but often misunderstood example being that of oxygen whose pressure (fugacity) can always be defined even when no gas phase is present. Components are preserved during reactions and phase changes, and their abundances are independent of the physical conditions in which the system finds itself. Species are all the forms of chemical associations of components, whether present or not (expressed or virtual) in the system. A species may be a $KAlSi_3O_8$ feldspar in a rock or an HCO_3^- ion in a solution. Species abundances are altered by temperature, reactions, and phase changes.

2.1 Conservation of mass

Let us begin with a few simple questions of geochemistry.

1. A river receives the input of a tributary. Given the flow rate and chemical composition of the two streams ahead of their confluence, what is the composition of the river water downstream of the confluence?
2. What is the chemical composition of a sediment for which the composition and abundance of the clay and quartz fragments are known?
3. A certain percentage of olivine of known composition precipitates from a basalt lava. What is the composition of the residual melt?

All three questions draw on the same principle: the whole is the sum of its parts. The elements of the river downstream are a combination of all of the upstream elements; the rock combines all of the elements present in the minerals; the parent basalt can be reconstructed by reincorporating the olivine in the residual lava. This principle is perfectly applicable, provided simply that all the components forming the aggregate system have been identified for certain. It is applicable whether the initial components have lost their physical identity during mixing (case 1) or whether their identity has been preserved, as in sand (case 2).

Let us try to determine the silicon content of the sediment in question 2. We have a mass M_{sed} of sediment composed exclusively of M_{clay} kilograms of clay and M_{qz} kilograms of quartz. The concentration of silicon in the clay fraction is C_{clay}^{Si}, that of the quartz fraction C_{qz}^{Si}, and that of the total sediment C_{sed}^{Si}. The applicable conservation equations are first that of the total mass of material:

$$M_{sed} = M_{clay} + M_{qz} \tag{2.1}$$

and that of silicon:

$$M_{sed} C_{sed}^{Si} = M_{clay} C_{clay}^{Si} + M_{qz} C_{qz}^{Si} \tag{2.2}$$

Dividing (2.2) by (2.1) yields:

$$C_{sed}^{Si} = f_{clay} C_{clay}^{Si} + \left(1 - f_{clay}\right) C_{qz}^{Si} \tag{2.3}$$

where the weight fraction, $f_{clay} = M_{clay}/M_{sed}$, of the clay component is shown. Notice that the conservation of mass has been written in the trivial form $f_{clay} + f_{qz} = 1$. A similar expression may be obtained for each element, e.g. for aluminum:

$$C_{sed}^{Al} = f_{clay} C_{clay}^{Al} + \left(1 - f_{clay}\right) C_{qz}^{Al} \tag{2.4}$$

Subtracting the concentration of each corresponding element in the quartz from both sides of (2.3) and (2.4) and dividing one by the other gives a single equation no longer dependent on f_{clay}. This indicates that

Figure 2.1: Graphic rules for conservation of mass and for mixing. A mineral or rock composition can be represented by a vector in the chemical concentration space (here Al and Si). Points 1, 2, and 3 represent three compositions of mixtures of clay and quartz. Notice the linear character of the mixing relationships.

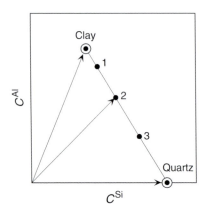

in a $\left(C_{\text{sed}}^{\text{Al}}, C_{\text{sed}}^{\text{Si}}\right)$ diagram (Fig. 2.1), the silicon and aluminum concentrations of sediments obtained by varying the proportion of quartz and clay form a straight line (mixing line) through the points representing the compositions of the two components. A mixture like this with linear properties is said to be conservative.

More generally, we are interested in a component j of mass M_j and an element i whose concentration in the component j is C_j^i. Here, i might refer to silicon and j to, say, the clay content of the sediment. The properties of the entire system, here the sediment, can be denoted by index 0, and the conservation equations can be written in the form:

$$\sum_j f_j = 1 \tag{2.5}$$

$$C_0^i = \sum_j f_j C_j^i = 1 \tag{2.6}$$

Equation 2.5 is known as the closure condition and indicates that the list of constituents is exhaustive. Equation 2.6 indicates that the concentration of the system as a whole is the weighted mean concentration of the composition of its components. The fraction by mass of each component in the mixture is the "weight" by which the concentration of each component must be multiplied to represent its contribution to the total inventory of the element in the entire system. An example of this principle, taken from everyday life, is the simple concept of rock. A rock is a sample, typically a piece of boulder or cliff, from what is actually no more than knocked off by the geologist's hammer. There is no such pure thing as a rock in nature: it is an artefact, whose composition results from a chance combination of all crystals collected by that particular hammer blow.

Vectors provide a useful means of describing the relationships of conservation. A mineral, a solution, or a rock in which the concentrations of n elements ($i = 1, \ldots , n$) have been measured are represented by a point (or a vector) in an n-dimensional space. In the example above, a vector equation could be written formally in the two-dimensional space of Si and Al concentrations:

$$\begin{pmatrix} C_{\text{sed}}^{\text{Si}} \\ C_{\text{sed}}^{\text{Al}} \end{pmatrix} = f_{\text{clay}} \begin{pmatrix} C_{\text{clay}}^{\text{Si}} \\ C_{\text{clay}}^{\text{Al}} \end{pmatrix} + f_{\text{qz}} \begin{pmatrix} C_{\text{qz}}^{\text{Si}} \\ C_{\text{qz}}^{\text{Al}} \end{pmatrix} \tag{2.7}$$

supplemented, of course, by the closure condition $f_{\text{clay}} + f_{\text{qz}} = 1$. Equation 2.7 shows the sediment vector as a linear combination of the vectors representing the components. If the sediment is composed of three components (let us add, for the sake of illustration, a carbonate component), then a vector formed by the concentration of the three elements in the rock will lie within the triangle formed by the equivalent vectors representing the clay, quartz, and carbonate components. Generally, when trying to interpret the chemical composition of a mixture in terms of constituents that are sometimes too numerous to be shown graphically, statistical methods are used, the simplest and most effective of which is principal component analysis (PCA).

Representing compositions in vector form also sheds light on the concept of reaction assemblage, or mineral reaction, that is so useful in metamorphic petrology. Figure 2.2 shows a natural assemblage of minerals where pyroxene and garnet have reacted to produce amphibole and corundum.

Let us illustrate these principles through a simpler mineral assemblage, i.e. the quartz (SiO_2), forsterite (Mg_2SiO_4), and enstatite ($Mg_2Si_2O_6$) assemblage. A rock composed of two components or, as is customary in petrology, two oxides, cannot have more than two minerals

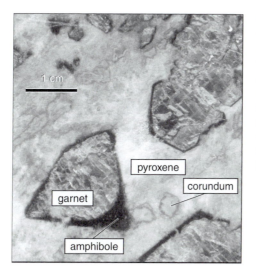

Figure 2.2: In this ancient metamorphosed basalt (eclogite), the fall in pressure when the rock rose to the surface destabilized the mineral assemblage. Pyroxene and garnet reacted with fluids no longer visible to give amphibole and corundum (sample courtesy of P. Thomas).

Figure 2.3: There cannot be three independent vectors in a two-dimensional space. A third mineral, therefore, requires there to be a reaction. Here forsterite + quartz ⇔ enstatite. This eliminates the deficient mineral, either quartz or forsterite.

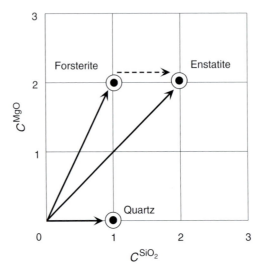

co-existing at equilibrium. Three minerals cannot be stable simultaneously because, in (SiO₂, MgO) space, any "rock" vector could be represented by an infinite number of combinations of "mineral" vectors whose equivalence is represented by the reaction forsterite + quartz ⇔ enstatite (Fig. 2.3). Thermodynamics tells us that this reaction evolves spontaneously from left to right, causing complete consumption of either quartz or forsterite. The stable mineralogical assemblage therefore includes only two minerals: the enstatite and, depending on the composition of the rock, either the quartz or the forsterite.

Geochemists are fond of ratios of elements or isotopes. There are two reasons for this: ratios are generally measured more precisely than concentrations and they are less sensitive to dilution phenomena. For example, the abundance of olivine in a basalt does not alter the ratio of those elements that are segregated in the melt (incompatible elements). The evaporation of seawater concentrates Cl^-, Na^+, ions, etc., but does not alter their concentration ratios. It is a serious mistake to try to apply the rules for absolute concentrations to ratios. The results stated below are explained in Appendix B. For the ratio A/B (e.g. the Si/Al ratio), the conservation equations become:

$$\sum_{\text{compos. } j} \varphi_j^B = 1 \tag{2.8}$$

$$\left(\frac{C^A}{C^B}\right)_0 = \sum_j \varphi_j^B \left(\frac{C^A}{C^B}\right)_j \tag{2.9}$$

where the weight $\varphi_j^B = f_j C_j^B / C_0^B$ represents the fraction of element B residing in phase j. Equation 2.8 is a closure condition equivalent to

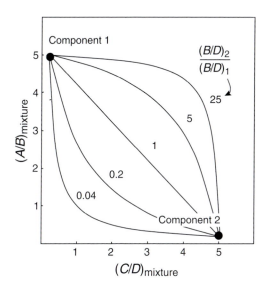

Figure 2.4: When concentration ratios or isotope ratios are plotted, the mixture or fractionation are not generally given by straight lines but by hyperbolas whose concavity depends on the abundance ratios of the elements in the denominator in the components of the mixture.

(2.5), but this time restricted to element B. Let us now compare the combination of ratios in a mixing process using (2.9). It can be seen that, unlike the weights f_j assigned to the concentration balance (2.6), weights φ_j^B assigned to the ratio balance are dependent on the element in the denominator of the ratio. Therefore, in ratio/ratio diagrams, mixtures do not generally form straight lines but hyperbolas (Fig. 2.4). A mixture relation like this is linear only if the denominator element of each ratio is the same (or in a constant ratio). This is why the conservation relations in ternary petrological diagrams, whose coordinates are normalized to the sum of the three variables depicted (as in the famous AFM plots), are linear. By contrast, plots involving fractional parameters, such as the magnesium number mg# = Mg/(Fe + Mg), which is widely used in petrology, must be handled with care.

2.2 Elemental fractionation

It is very useful to introduce partition coefficients when studying the substitution of minor elements or trace elements in the lattice of minerals in equilibrium with magmatic fluids or natural solutions from which they precipitate. When an element i is in solution in two co-existing phases j and J (e.g. j stands for seawater and J for a carbonate that precipitates out), the Nernst law can be written:

$$\frac{x_J^i}{x_j^i} = K(T, P, x) = k_0 \exp\left(-\frac{\Delta G_0}{RT}\right) \tag{2.10}$$

where x_j^i is the molar proportion of element i in phase j, R is the gas-law constant, and ΔG_0 a measure of the energy of exchange of this element between the two phases j and J. The partition (or distribution) coefficient K depends on the temperature T, pressure P, and composition of the phases. The pre-exponential factor k_0 is a measure of the non-ideality of the solutions. After a simple adjustment that takes molecular weights into account, the concentration ratios between phases may also be described by partition coefficients. For only slightly concentrated elements (trace elements), the ratio of concentration of an element between two phases under comparable conditions of temperature, pressure, and aggregate composition of the system (acid, basic, aqueous) is constant. It is a common usage in geochemistry to restrict the term partition coefficients to mineral/liquid coefficients, but this choice is arbitrary and occasionally misleading. A liquid/mineral partition coefficient is a perfectly valid concept. Figure 2.5 gives a few examples of mineral/liquid partition coefficients in magmas.

The term "reasonably" justifies the activity of many experimental and analytical laboratories. To a fairly good degree of approxima-tion, the dependence of K on temperature and pressure can be des-cribed by:

$$\mathrm{d}\ln K = \frac{\Delta H}{RT^2}\mathrm{d}T + \frac{\Delta V}{RT}\mathrm{d}P \qquad (2.11)$$

(Appendix C) where ΔH and ΔV measure differences in the enthalpy and molar volume of element i between the two phases. The depen-dence of K on composition is more complex: a satisfactory model for

Figure 2.5: Typical partition coefficients for some important trace elements between the main minerals and the liquid for a basalt composition.

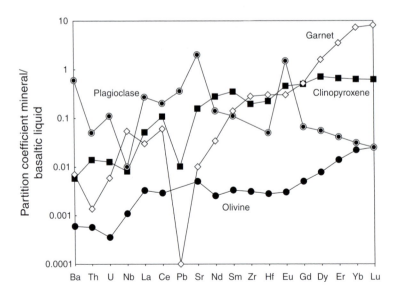

solids is Onuma's elastic model, in which energies are elastic energies required by the substituting element to replace that element that normally makes up the mineral. In order to insert an atom of ionic radius r_0 into a crystallographic site, which we consider as a spherical cavity of radius r, work must be done against the electrostatic forces F. As a first approximation, these forces can be taken as proportional to the contraction or expansion of the ionic radius from r to r_0, i.e. $F \approx k(r - r_0)$ (Hooke's law). In this equation, k is a constant related to certain elastic properties of the medium known as Young's modulus and the Poisson ratio. A measure of the change in elastic energy U upon compression or expansion is:

$$dU = -PdV = -\frac{F}{4\pi r^2} 4\pi r^2 dr = -F dr \qquad (2.12)$$

which is integrated between r_0 and r as:

$$\Delta U = -\frac{k}{2}(r - r_0)^2 \qquad (2.13)$$

Most ions of identical charge fit in with such a parabolic relationship between the binding energy, and therefore $\ln K$, and their squared radius (Fig. 2.6). Measurement of the different ΔH values and evaluation of elastic energies are daunting tasks. Although fractionation theory applies remarkably well to both low- and high-temperature systems, it is essential to remember that it requires approximations, however good they may be, and that it cannot be compared with a zero-order principle such as the conservation of mass.

    If one phase plays a particular role, such as seawater or a magmatic liquid from which minerals precipitate, the fractionation of elements can be described by taking this phase as a reference and introducing an

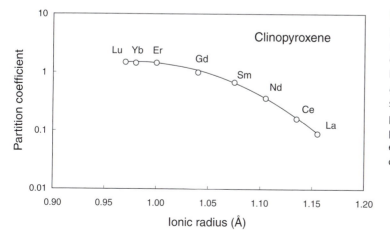

Figure 2.6: Variation of the partition coefficients of rare-earth elements between clinopyroxene and basaltic liquid by ion radius (Blundy et al., 1998). The parabolic shape of the curve reflects a physical control of the partition coefficient by the elastic properties of the crystalline lattice.

aggregate (bulk) solid/liquid partition coefficient $D^i_{s/l}$ such as:

$$\sum_{\text{solids } j} f_j = 1 \qquad (2.14)$$

$$D^i_{s/l} = \sum_{\text{solids } j} f_j K^i_{j/\text{liq}} \qquad (2.15)$$

described as the weighted mean of individual partition coefficients $K^i_{j/\text{liq}}$ of element i between mineral j and the liquid, the weights being the abundance of each mineral in the precipitate. Although (2.15) is exact, this equation does not suppose that $D^i_{s/l}$ remains constant when the mineral fractions f_j change during the process. Later we will come across geological settings where these conditions are fulfilled. $D^i_{s/l}$ provides a measure of the element compatibility (or of its refractory character) in a given environment, and it is standard practice in igneous geochemistry to represent the concentrations of elements of different magmatic rocks in order of increasing compatibility as the mantle melts, the mean order being established by experiment. These "spidergrams" entail normalization, i.e. division by concentrations of a reference rock or reservoir (chondrites, Bulk Silicate Earth, mid-ocean ridge basalts = MORB) and have become a standard graphic tool in geochemistry (Fig. 2.7).

Let us apply the simple theory of chemical fractionation to the change in the concentration of an element i during partial melting of a source rock made up of several minerals. This theory leads to the so-called equilibrium- or batch-melting equations. If the proportion of liquid in

Figure 2.7: Spidergram of trace elements in essential types of basalts standardized to a model of concentration of the primitive mantle. The incompatibility of elements, i.e. the tendency of crystals to reject them to the melt, diminishes toward the right. MORBs are "depleted" in incompatible elements, while ocean island basalts (OIB) are "enriched" relative to "primitive mantle."

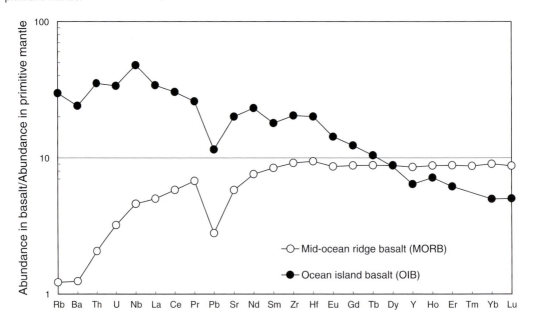

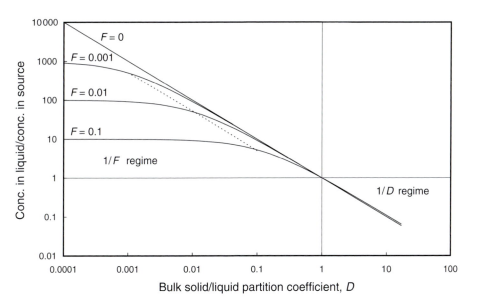

the molten rock is F (melt fraction), the mass balance is written:

$$C^i_{source} = FC^i_{liq} + (1-F) \sum_{solids\ j} f_j C^i_j = \left[F + (1-F)D^i_{s/l}\right] C^i_{liq} \quad (2.16)$$

or

$$C^i_{liq} = \frac{C^i_{source}}{F + (1-F)D^i_{s/l}} \quad (2.17)$$

This equation separates two regimes of melting that are best understood by plotting the enrichment factor C^i_{liq}/C^i_{source} during melting as a function of the bulk solid/liquid partition coefficient (Fig. 2.8). For $F < D^i_{s/l}$, the enrichment factor is nearly constant and equal to $1/D^i_{s/l}$, while for $F > D^i_{s/l}$, the enrichment factor varies approximately as $1/F$. In the mantle or crust, the melt fraction F is generally low, of the order of 0.1 to 15%. For incompatible elements such as Th, Nb, La, or Ba, $D^i_{s/l}$ is virtually negligible and the elements are very rapidly extracted from the residue: their concentration in the liquid is inversely proportional to the melt fraction and so varies very rapidly at low values of F. For a highly compatible element like nickel, $D^i_{s/l}$ is very high, while F remains low compared with unity. For the compatible elements, concentration C^i_{liq} varies very little with F. Concentrations of primary basalt liquids in nickel, but also in magnesium, chromium, and osmium, are therefore buffered by the melt residue. Buffering of elements with large $D^i_{s/l}$ by the solid residue is clearly visible in Fig. 2.9, which shows that the variability of basalt compositions falls very markedly as the level of compatibility of the element increases.

Figure 2.8: Enrichment/depletion factor during melting as a function of the bulk solid/liquid partition coefficient D for different degrees F of melting. Note the two regimes separated by the dashed line $F = D$: for $F < D$, the enrichment factor remains constant and equal to $1/D$; while for $F > D$, it varies as $1/F$. Efficient fractionation between very incompatible elements requires small melt fractions.

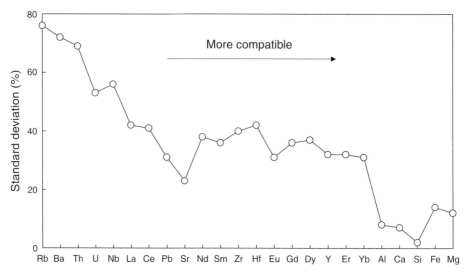

Figure 2.9: The variability of concentrations of elements in basalts (measured by the relative standard deviation) decreases with their compatibility (after Hofmann, 1988). This reflects the buffering of the more compatible element concentrations in melts by the residual solid.

However complex the distributions observed in rocks, and especially in magmatic rocks, it is useful to understand that certain elements are the unmistakable fingerprint of certain minerals. For example, olivine concentrates Ni, garnet concentrates heavy rare-earth elements, the pyroxenes concentrate Sc and Cr, and the feldspars concentrate Eu and Sr. The trained observer can quickly spot from an analysis of trace elements the influence of certain minerals during the formation of a rock.

Separate treatment is reserved for elements that reach the limit of saturation leading to phase changes. This is what happens with zirconium in felsic lavas and granites when zircon ($ZrSiO_4$) precipitates, with chromium in basalts whenever chromite forms, and more importantly still with water or CO_2, which separate out in a vapor phase when magmas ascend. At this point, the concentration of the element forming the major phase (e.g. Zr or CO_2) remains fixed, we say buffered, by either the zircon or the vapor. The balance equations have to be rewritten to cover this new situation.

2.3 Isotopic fractionation

Isotopic, thermodynamic, or kinetic modes of fractionation are referred to as mass-dependent processes, in contrast with the variations engendered by radioactivity or during processes of formation of elements in the stars considered later. The logic underlying thermodynamic fractionation can be applied to isotopes with the added advantage that the behavior of two isotopes of the same element will be more similar than that of two distinct elements, and, indeed, the chemistry of two isotopes of a heavy element, e.g. lead, is virtually unchanged. The applications of isotopic fractionation to the mapping of low- and medium-temperature phenomena are fundamental. The principles are rather similar to those set out above for chemical fractionation. Let us imagine how isotopes ^{16}O ($\approx 99.8\%$ of natural oxygen) and ^{18}O are distributed between liquid water and vapor. The reaction can be written:

$$H_2{}^{18}O_{liq} + H_2{}^{16}O_{vap} \Leftrightarrow H_2{}^{16}O_{liq} + H_2{}^{18}O_{vap} \qquad (2.18)$$

The mass action law governing this equilibrium is written:

$$\left(\frac{H_2{}^{18}O}{H_2{}^{16}O}\right)_{vap} \Big/ \left(\frac{H_2{}^{18}O}{H_2{}^{16}O}\right)_{liq} = \left(\frac{^{18}O}{^{16}O}\right)_{vap} \Big/ \left(\frac{^{18}O}{^{16}O}\right)_{liq} = \alpha^O_{vap/liq}(T, P) \quad (2.19)$$

The fractionation coefficient $\alpha^O_{vap/liq}(T, P)$ replaces the partition coefficient K to signal that the exchange concerns just a single nuclide. Fractionation coefficients are thus defined for many stable isotope pairs, such as deuterium/hydrogen (D/H), $^{13}C/^{12}C$, $^{15}N/^{14}N$, $^{34}S/^{32}S$, etc., between such varied phases as gases, natural solutions, magmatic liquids, and minerals. The isotopes exchanged have the same outer electron configuration. Their chemical properties are therefore very similar and their molar volumes almost identical, making α very close to unity and virtually independent of pressure.

The energy of a chemical bond, e.g. H—H for the diatomic hydrogen molecule, varies with the distance between the bonding atoms (Fig. 2.10). Energetically, the most favorable position of the electrons is between the nuclei, which leads to a mutual attraction of the two hydrogen atoms. Pairing electrons on a molecular orbital results in a net gain of energy; but when the atoms get too close, interaction between their electronic shells produces a repulsive potential. These competing effects tend to confine the two hydrogen atoms at an optimum distance at the lowest point of the potential "well." Quantum mechanics, however, requires that the energy of a particular chemical bond takes discrete (quantized) values. The lowest possible energy (ground state) of the bond is higher than the minimum of the potential well: the difference between the ground state and the minimum on the potential curve is called the zero-point energy. Stable-isotope fractionation results from the existence of the zero-point energy.

With the discovery of the photo-electric effect by Einstein, we learned that the energy associated with a periodic movement, such as thermal vibrations and electromagnetic waves, is proportional to the vibration frequency. Heavy bodies react more slowly than light ones and therefore tend to occupy lower energy levels. Bond energy varies with $1/\sqrt{M}$, where M is the mean mass of the atoms that form the molecule. A consequence of this rule is that the quantized energy levels are lower

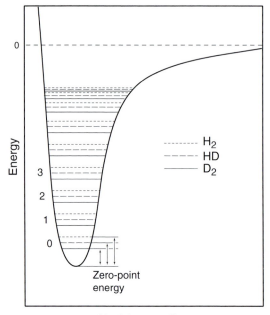

Figure 2.10: Energy of the hydrogen−hydrogen bond. The potential of the H−H bond as a function of the distance between the two nuclei is shown as a continuous curve. The quantized energy levels ($n = 0, 1, 2, 3, \ldots$) are shown for the H−H (hydrogen−hydrogen), H−D (hydrogen–deuterium), and D−D (deuterium−deuterium) bonds. The first energy levels are nearly equally spaced. The isotopically heavier molecules are in a lower state of energy than the lighter molecules. Zero-point energy is the elevation of the ground state ($n = 0$) above the minimum of the potential well. Differences in zero-point energies between molecules with different isotopes (e.g. H−H and H−D) account for all stable-isotope fractionation effects.

for a liaison involving the heavier isotopes: the molecule D_2 is more stable than H_2 (Fig. 2.10). Since transforming liquid into vapor requires addition of energy to break bonds (latent heat), vapor offers sites that are higher on the energy scale than the corresponding liquid and therefore concentrates the more energy-hungry lighter isotopes. Solids and liquids, where most of the energy is stored in vibrations, lower the total energy of the system by concentrating heavier isotopes. In a liquid–vapor equilibrium, such as H_2O liquid and vapor, the liquid is enriched in the heavier isotope (e.g. ^{18}O), while the vapor preferentially concentrates the lighter isotope (e.g. ^{16}O).

Temperature, which, through the Boltzmann distribution, determines the population of atoms whose energy exceeds a given threshold, plays an important role in the stability of bonds, and hence in isotopic

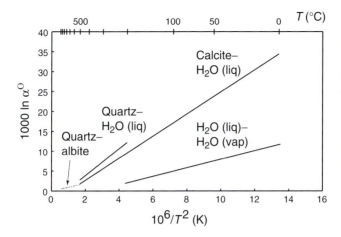

Figure 2.11: The fractionation of oxygen isotopes between a mineral phase and water, or between different mineral phases. Notice that preferential incorporation of ^{18}O is normally favored by the mineral structure and that isotopic fractionation decreases rapidly with increasing temperature.

fractionation. At ambient temperature and above, we observe (Fig. 2.11) fractionation laws of the type:

$$\ln \alpha^O = \frac{A}{T^2} + C \qquad (2.20)$$

where A and C are constants. These laws are widely used for thermometry of magmatic, metamorphic, or hydrothermal rocks. Notice the variation in $1/T^2$, which is a quantum effect specific to isotope exchanges at medium to high temperatures (a $1/T$ dependence would prevail at low temperature). The symmetry constant C is normally zero for isotopic exchange between anhydrous minerals so that fractionation at magmatic temperatures is normally negligible.

For elements with more than two isotopes, at constant temperature, the amplitude of fractionation α increases with the difference in mass of the isotopic ratios. For example, the difference in the $^{18}O/^{16}O$ ratio between two samples is twice the difference in the $^{17}O/^{16}O$ ratio. This is because bond energy varies with the mass of the bonded atoms. Therefore $\ln \alpha$ can be expanded to the first order relative to the difference in mass Δm in the isotope ratio, giving:

$$\ln \alpha (\Delta m) = \ln \alpha (\Delta m = 0) + f \Delta m + \mathcal{O} (\Delta^2 m) \qquad (2.21)$$

where f, which is the derivative of $\ln \alpha$ relative to Δm for $\Delta m = 0$, is a coefficient that is independent of mass, termed mass discrimination. In this equation, $\mathcal{O}$ means "of the order of" and we will neglect terms of higher than first order. Moreover, when the mass difference is zero, there is no fractionation between a mass and itself ($\alpha = 1$), and the first term on the right-hand side is therefore zero, giving for fractionation of

^{18}O and ^{16}O between phases 1 and 2:

$$\alpha^O_{2/1} = \left(\frac{^{18}O}{^{16}O}\right)_2 \Big/ \left(\frac{^{18}O}{^{16}O}\right)_1 = e^{f\Delta m} \tag{2.22}$$

When the fractionation is small, i.e. when f tends toward 0, we utilize the linear approximation obtained by developing the logarithm of the left-hand side of (2.21) to the first order:

$$\alpha^O_{2/1} \approx 1 + f\Delta m \tag{2.23}$$

We will have, for example, the two equations:

$$\left(\frac{^{17}O}{^{16}O}\right)_2 = \left(\frac{^{17}O}{^{16}O}\right)_1 (1+1f) \tag{2.24}$$

$$\left(\frac{^{18}O}{^{16}O}\right)_2 = \left(\frac{^{18}O}{^{16}O}\right)_1 (1+2f) \tag{2.25}$$

It will be seen later that one very important application of this simple theory is in the classification of planetary bodies.

Let us now introduce delta notation, which is widely used for isotopic ratios of stable nuclides. $\delta^{18}O$ and $\delta^{17}O$ represent the deviations of the isotopic ratios in the sample in parts per thousand relative to the same ratio in a reference material. For oxygen, the reference material is either the standard mean ocean water (SMOW) or a belemnite from the Pee Dee formation (PDB). We have:

$$\delta^{18}O = \left[\frac{(^{18}O/^{16}O)_{sample}}{(^{18}O/^{16}O)_{ref}} - 1\right] \times 1000 \tag{2.26}$$

A very useful approximation of the fractionation properties is to compare the difference Δ of $\delta^{18}O$ with the fractionation coefficients α (beware of the phase order). The following notation is used:

$$\Delta_{2-1} = \delta^{18}O_2 - \delta^{18}O_1 \approx 1000 \ln \alpha \tag{2.27}$$

If we choose phase 1 as a reference in (2.24) and (2.25), it can be seen that $\delta^{17}O = 1000f$ and $\delta^{18}O = 2000f$. A plot of $\delta^{17}O$ versus $\delta^{18}O$ clearly shows that all samples from the Earth and the Moon form a single alignment (Fig. 2.12) with a slope of 0.5, termed the mass-fractionation line. It can be inferred that the Earth and Moon formed from the same material with a uniform oxygen isotope composition. This argument is central to the theory whereby the Moon originated by a minor proto-planet colliding with the Earth. However, the other meteorites form different groups, showing that they formed in different regions of the solar nebula.

Since bond energy varies with the inverse square root of the mean mass of the atoms that form the molecule, the thermodynamic fractionation factor f between isotopes generally falls very quickly with

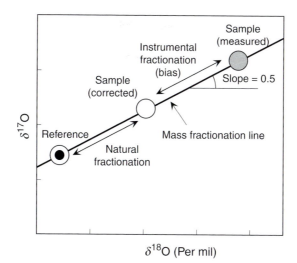

$\delta^{18}O$ (Per mil)

Figure 2.12: Fractionation depending on the mass of isotopes ^{16}O, ^{17}O, and ^{18}O is consistent whether its cause is natural or analytical (see also Fig. 9.5). The $^{17}O/^{16}O$ and $^{18}O/^{16}O$ ratios are shown in delta notation. All samples from Earth lie on the same straight line of mass fractionation whose slope is equal to the ratio of the mass difference $(17 - 16)$ by difference $(18 - 16)$, i.e. 0.5. Natural fractionation is a deviation determined relative to an arbitrary reference sample, hence the δ notation. The isotopic bias is introduced by the mass spectrometric analysis and follows a similar mass dependence law: it must be corrected for each sample by appropriate techniques.

increasing mean atomic mass as it varies with the difference between smaller and smaller quantities. Fractionation extents in the range 2–8% are commonly observed for the D/H ratio, with a figure in the range 0.2–4% for oxygen isotopes, and falling to 0.1% for zinc isotopes. It is important to understand, at this stage, that isotope measurements made in the laboratory report isotopic fractionation values that are the sum of fractionation related to the action of all natural processes *and* of instrumental fractionation, sometimes referred to as instrumental mass bias (Fig. 2.12). We have seen that in a condensed state, either solid or liquid, the bond of the heavier isotope with its surrounding atoms is more stable than that of a light isotope. An important consequence of this is that the lighter isotope (^{1}H, ^{16}O) is usually more abundant than the corresponding heavy isotope (D or ^{2}H, ^{18}O) in water vapor, compared with liquid water. A second consequence is the distribution of oxygen isotopes in the natural environment (Fig. 2.13). Water at low and medium temperature tends to concentrate the light isotope relative to silicate minerals at equilibrium with it: the Earth has a mean $\delta^{18}O$ of $+5.5$ per mil relative to SMOW, which is the value of mantle peridotites, the main reservoir of this element. Under the influence of low-temperature weathering and

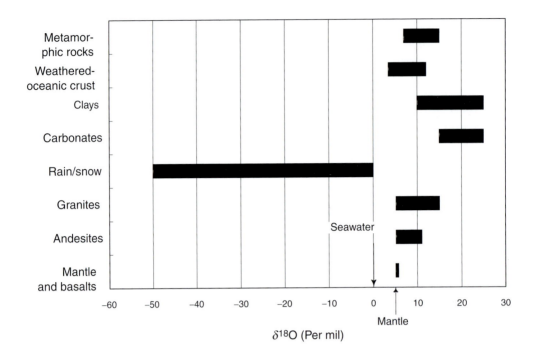

Figure 2.13: Distribution of $\delta^{18}O$ values in rocks and natural water. Notice seawater at 0 per mil and the Earth's mantle at 5.2 per mil. It is the equilibration of sedimentary and metamorphic rocks with seawater at low temperature that raises the $\delta^{18}O$ of these rocks and lowers that of seawater. The very great isotopic variation of meteoric water results from atmospheric precipitation which occurs at low temperature.

submarine hydrothermal reactions, heavy isotopes concentrate in clays and carbonates at equilibrium with seawater, which attain $\delta^{18}O$ values of $+12$ to $+25$ per mil. As a result of the preferential incorporation of ^{18}O in low-temperature sedimentary minerals, seawater acquires a $\delta^{18}O$ of 0 per mil, lower than the mean value on Earth. Marine alteration, especially the interaction of seawater with mid-ocean ridges, is thought to be responsible for the current $^{18}O/^{16}O$ ratio of seawater: throughout the Earth's history, seawater has circulated in submarine hydrothermal systems and its oxygen has been brought into isotopic equilibrium at temperatures of the order of 275 °C with the oxygen of basalts erupted by the volcanic systems of mid-ocean ridges.

The commonest rocks in the continental crust derive largely from the metamorphic transformation of sediments (schist, gneiss) and their melting (granite). Although interaction between crustal rocks and the fluids percolating through the crust may affect the ultimate isotopic composition of their oxygen, the high $\delta^{18}O$ (typically $+7$ to $+15$ per mil) of gneiss, mica-schist, and granite reflects this so-called meta-sedimentary origin. Such high $\delta^{18}O$ values of metamorphic and magmatic rocks with respect to the mantle value ($+5.5$ per mil), require the presence in their source material of rocks and minerals that have been subjected to low-temperature interaction with a hydrous fluid, such as seawater, ground water, or meteoritic water.

The ^{12}C and ^{13}C isotopes fractionate intensely during exchanges between reduced carbon (dissolved in silicates and magmatic carbonates) and gas molecules, especially when these fractionation events imply very large differences in carbon configuration symmetry between the other atoms of carbonate species (CO_3^{2-}, HCO_3^-, CO_2, CH_4). Such fractionation is also highly dependent on temperature, and the reactions between solids and the different gases must be considered. The same would apply to $^{34}S/^{32}S$ fractionation between the different species containing sulfur (especially S^{2-} and SO_4^{2-}). These complex processes lie beyond the scope of this book, but it is worth noting that, because of the sensitivity of isotopic fractionation to conditions of pH and oxidation–reduction (redox) reactions, these isotopic systems are particularly significant for the genesis of ore bodies.

The fractionations just referred to concern the distribution of isotopes between phases in equilibrium, but another type of fractionation (kinetic) occurs when biological processes are involved. For example, the carbonic anhydrase enzyme found in mammal blood speeds up the conversion of CO_2 into an HCO_3^- ion by five orders of magnitude, thus preventing bubbles of respiratory CO_2 from forming. For high reaction rates, equilibrium cannot be achieved and the light isotopes, requiring lower transfer energies, are exchanged more readily. Thus, by comparison with the environment within which they formed, organic products are significantly enriched in ^{12}C relative to ^{13}C and in ^{32}S relative to ^{34}S. Such isotopes are extensively used in studying the genesis of fossil fuels and low-temperature mineralization in which biological processes are important.

Such large extents of kinetic fractionation are of considerable importance for carbon bio-geochemistry. The isotopic standard of this element is a marine carbonate (PDB is a belemnite from the Pee Dee formation). Marine carbonates, as a whole, will therefore have $\delta^{13}C$ values close to 0 per mil. The carbon in atmospheric CO_2 has a $\delta^{13}C$ value close to -7 per mil. This difference arises from normal thermodynamic fractionation. However, the continental biomass (dominated by plants) has a distribution with two maxima, one at about -14 per mil, corresponding generally to grasses, and a larger one at about -25 per mil, corresponding to most broad-leaved plants (Fig. 2.14). This contrasted fractionation originates in photosynthesis and does not take place at equilibrium. It is brought about by two different CO_2 fixation mechanisms by very different plants, with what are known as the C4 cycle, predominating for grasses, and the C3 cycle, for broad-leaved plants and conifers. The variation in $\delta^{13}C$ of fossil plant matter collected from bores in peat bogs is thus utilized for tracing changes in vegetation between glacial and interglacial periods. Increased $\delta^{13}C$ levels of organic matter are interpreted

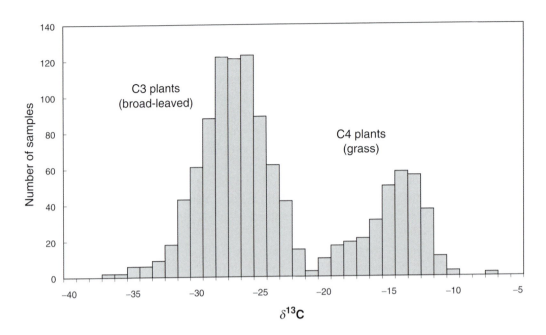

Figure 2.14: The effect of the metabolic cycle on the fractionation of carbon isotopes during photosynthesis. The C3 cycle (broad-leaved plants) and C4 cycle (grasses) correspond to very different photosynthetic processes within the cell. Past climatic conditions can therefore be inferred from the $\delta^{13}C$ value of organic remains (Deines, 1980).

as marking the advance of grasslands, while very negative $\delta^{13}C$ levels indicate a return of broad-leaved plants. Phytoplankton, which dominate primary production at the surface of the oceans, have a $\delta^{13}C$ value of about -25 per mil. Where organic matter production is intense, there is a corresponding depletion of surface water in ^{12}C. The carbon of carbonates that form the tests of calcareous algae (such as coccoliths that form chalk, in particular) and of foraminifera that graze on plankton (such as globigerina) reflects the intensity of this depletion. The $\delta^{13}C$ values of foraminifera in carbonate sediments serve as a measuring rod for the biological productivity of ancient oceans where the sedimentation occurred.

The isotopes of nitrogen, ^{14}N and ^{15}N, are utilized for special applications such as bio-geochemistry, cosmochemistry, and diamond geochemistry. The depletion of the light isotope ^{14}N of nitrate dissolved in seawater by photosynthetic activity is very important. Just as for $\delta^{13}C$, there is a very substantial increase in the $\delta^{15}N$ values of nitrates in surface waters compared with those of deep waters.

It is finally worth noting, too, that fractionation independent of mass has recently been observed in atmospheric oxygen in conjunction with self-catalyzing processes and the ozone cycle. The processes that give rise to these isotopic effects are not, so far, as well understood as mass-dependent fractionation.

2.4 Distillation processes

The distillation of alcohol, whereby we seek to enrich the ethyl-alcohol content of the condensate resulting from the boiling of wine or any other low-grade fermented beverage, is a familiar enough example. When a finite quantity of a substance changes state and the product of the transformation is isolated, the progressive chemical or isotopic fractionation associated with the progress of this transformation defines the phenomenon of distillation. This process occurs naturally in various geological settings:

- fractional crystallization of magmas during which solids, termed cumulates, are successively isolated from the residual magma;
- fractional melting of the mantle, producing liquids extracted immediately from the molten source;
- progressive condensation of atmospheric water vapor, in the course of which precipitation loses contact with the high atmosphere H_2O;
- boiling of hydrothermal solutions.

It can be shown (Appendix D) that the change in concentration C^i_{res} of an element i in the residual phase during formation of a new phase obeys Rayleigh's law:

$$\mathrm{d} \ln C^i_{res} = \left(D^i - 1\right) \mathrm{d} \ln f \qquad (2.28)$$

where D^i is the bulk partition coefficient between the new phase and the parent residual phase (the order is essential) and f the fraction by mass of the residual phase relative to the stock of original material. The most common form of this equation is:

$$C^i_{res} = C^i_0 f^{D^i - 1} \qquad (2.29)$$

where C^i_0 is the concentration in element i of the parent phase of the original material, i.e. for $f = 1$.

A first application of this theory is the fractional crystallization of magmas, for which we obtain:

$$C^i_{liq} = C^i_0 f^{D^i - 1} \qquad (2.30)$$

Here, parameter f represents the residual liquid fraction and C^i_0 the composition of the parent magma. The concentration of the solid at equilibrium with the liquid is obtained by multiplying (2.30) by the partition coefficient of the element. For incompatible elements, like Th, Ba, and La, the partition coefficients D^i, which measure solid/liquid fractionation, are low: their concentration in the residual magma and, consequently, in the extracted mineral phases is therefore inversely proportional to f. Over most of the range of magma differentiation, liquid

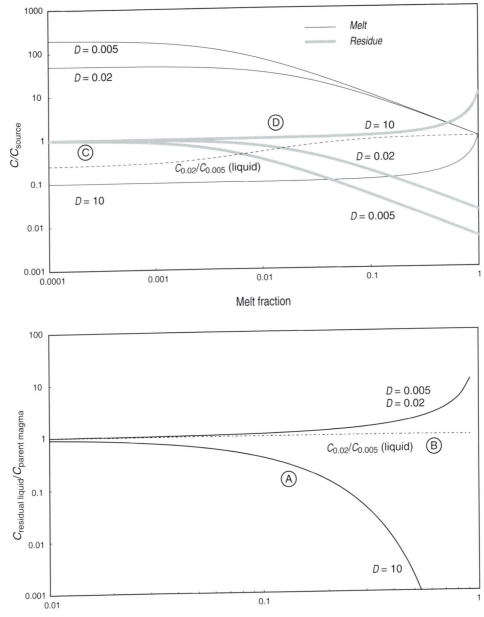

Figure 2.15: Comparison of the effect of fractional crystallization (bottom) and partial melting (top) on the concentration in magmatic melts of elements with different compatibilities. The diagrams show the case of bulk partition coefficients for elements with $D = 0.005$, 0.02, and 10. Curve A: Even the removal of a few percent of a mineral hosting a very compatible element, e.g. Ni in olivine ($D \approx 10$), changes concentration in the residual melt drastically, whereas this change is hardly noticeable with highly incompatible elements such as Th, Ba ($D \approx 0.005$), or La ($D \approx 0.02$). Curve B: Even extreme mineral fractionation does not change the ratio of two incompatible elements such as Th/La in the residual melt. Curve C: For small to moderate fractions of melt, compatible elements remain buffered by the solid. Curve D: In contrast, the ratio of two incompatible elements is fractionated until the fraction of melt exceeds the largest of the two partition coefficients. At this stage, the bulk of the incompatible elements has been transferred to the melt and their relative concentrations are proportional to their abundances in the source. The dashed lines labelled $C_{0.02}/C_{0.005}$ represent the ratio of two elements with partition coefficients of 0.02 and 0.005, respectively.

is the dominant phase. Parameter f is therefore close to unity and the concentration of incompatible elements varies little during crystallization. Moreover, the ratios of incompatible elements (e.g. Th/La) in differentiated lavas are virtually unaffected by fractional crystallization. By contrast, for compatible elements, such as Ni and Cr, high D^i values cause very high variations in concentration in the residual magma for low variations of f. Thus the precipitation of 10–20% olivine may remove most of the nickel initially present in the magma (Fig. 2.15).

Conversely, the coefficients of interest for the fractional melting in which the liquid is progressively drawn off will be the liquid/solid fractionation coefficients. This gives the equation:

$$C^i_{\text{liq}} = \frac{C^i_0}{D^i} (1 - F)^{\frac{1}{D^i} - 1} \qquad (2.31)$$

where $F = 1 - f$ is the liquid fraction extracted, but C^i_0 is now the concentration of the source (mantle, crust, etc.). In a system with a low degree of melting, which is the usual case in nature, the value of F is close to 0. By symmetry with the foregoing reasoning, it can be seen that the concentrations of incompatible elements, which have very high liquid/solid partition coefficients (note the reverse order of phases), will vary greatly with the degree of melting. As in the case of partial fusion at equilibrium, concentrations of compatible elements remain, however, buffered by the residual solid. A simple, but crucial, principle is inferred: melting is investigated with incompatible elements, crystallization with compatible elements (Fig. 2.15).

Dividing (2.28) written for A by the ratio of the same equation written for B, the variable f can be eliminated:

$$\frac{d \ln C^A_{\text{res}}}{d \ln C^B_{\text{res}}} = \frac{D^A - 1}{D^B - 1} \qquad (2.32)$$

In a logarithmic plot, the so-called log–log plot, $(\ln C^A, \ln C^B)$, concentrations of the different phases present therefore form two-by-two alignments. Examples of this type of relation abound in the geochemical literature (Table 2.1). The linear relation between the δD and $\delta^{18}O$ of rain and snow discovered by Dansgaard can be explained by a process of fractional condensation of water vapor upon migration of moist equatorial air toward the poles. The partition coefficients of (2.28) are replaced by the α fractionation coefficients of deuterium and ^{18}O between liquid and vapor phases, and the variations in the isotopic ratios are sufficiently small for the logarithm of the variables to be replaced in this expression by the variable itself.

Table 2.1: *Distillation processes*

Process[a]	Parent phase	Evolving phase	Partition coefficient	Environment
Crystallization	Liquid	Solid	Solid/liquid	
	Magma	Cumulate	Magma/cumulate	Magmatic
	Brine	Salt	Salt/brine	Evaporitic
Melting	Source rock	Magma	Liquid/solid	Magmatic
Evaporation	Solution	Vapor	Vapor/liquid	Hydrothermal
Condensation	Vapor	Liquid water	Liquid/vapor	Atmospheric

[a] Add "fractional" to the term describing the process.

References

Blundy, J. D., Robinson, J. A. C. and Wood, B. J. (1998) Heavy REE are compatible in clinopyroxene on the spinel therzolite solidus. *Earth Planet*. Sci. *Lett*., **160**, pp. 493–504.

Deines, P. (1980). The isotope composition of reduced organic carbon. In *Handbook of Environmental Isotope Geochemistry*, ed. P. Fritz and J. C. Fontes, Vol. 1, pp. 329–406. Amsterdam: Elsevier.

Hofmann, A. W. (1988) Chemical differentiation of the Earth: the relationship between mantle, continental crust, and oceanic crust. *Earth Planet. Sci. Lett*., **90**, pp. 297–314.

Chapter 3
Geochronology and radiogenic tracers

We have seen that radioactivity is not dependent on the chemical bonding of atoms, nor on temperature, nor on pressure. Radioactivity can be described as an event whose probability of occurrence is independent of time and depends only on the duration of measurement. The probability that a radioactive nuclide will decay per unit of time is noted λ. This probability, termed the decay constant, is specific to the radioactive nuclide under consideration. Radioactive decay, like incoming calls at a telephone exchange, is a prime example of a Poisson process, in which the number of events is proportional to the time over which the observation is made. In the absence of any other loss or gain, the proportion of parent atoms (or radioactive nuclides) disappearing per unit of time t is constant:

$$\frac{dP}{P\,dt} = -\lambda \tag{3.1}$$

For a number of parent atoms $P = P_0$ at time $t = 0$, this equation integrates as:

$$P = P(t) = P_0 e^{-\lambda t} \tag{3.2}$$

In this form, (3.2) is not generally useful for measuring ages. To determine the age of a system from the measurement of the number of parent atoms at the present time, we must also know P_0. The half-life $T^{1/2}$, which is the time it takes for half of the parent nuclides to decay, is $\ln 2/\lambda \approx 0.69/\lambda$. After five half-lives, 98.5% of the radioactive isotopes have decayed away, and 99.9% have decayed after eight half-lives. The product λP measures the number of decay events per unit time. It is commonly referred to as the activity of the radioactive nuclide P and noted $[P]$. The becquerel (Bq) is a unit equal to one decay event (count) per second. A liter of seawater has an activity of 12 Bq, mostly because of the potassium-40 and uranium naturally dissolved in it. The human body contains enough potassium-40 and carbon-14 to register an activity of 5000–10 000 Bq as perfectly natural in origin. A granite cobblestone normally produces a radioactivity of several thousand Bq.

For each parent atom, a daughter atom (or radiogenic nuclide) is created, usually of a single element, whose number can be noted D. In a closed system and for a stable daughter nuclide D, the number of parent and daughter atoms is constant, therefore:

$$D = D_0 + P_0 - P = D_0 + P\left(e^{\lambda t} - 1\right) \tag{3.3}$$

The term $P(e^{\lambda t} - 1)$ is a measure of the accumulation of the radiogenic nuclide during time t. Even if D and P are measured, this equation is no more a timing device than the previous one unless we know the number of daughter atoms D_0 at time $t = 0$. A simple case where this condition applies is when the initial number of daughter nuclides is small enough so as to be negligible. This approximation is fairly generally valid for the potassium–argon dating method and for uranium–lead dating of zircons described later.

In many other instances, we apply the principle of isotopic homogenization to remove the ambiguity arising from our ignorance of the initial state of the system. Isotopic homogenization results from the fact that the chemical properties of different isotopes from the same element are very similar, though as seen in the previous chapter not identical. To help our understanding, this principle is illustrated in Fig. 3.1 by a playful comparison. A high-walled yard with a tree in the center represents two crystalline sites with different energy levels. In the first case, we release a few dozen cats and dogs into the yard–tree system and we can well imagine that after some snappy movement among our elements, they will arrange themselves in appropriate sites, cats in the tree and dogs on the ground in the yard, under the tree. Cats and dogs are two different elements in competition for the same sites: they arrange themselves spontaneously so as to move to the most stable configuration! Any alternative configuration is either improbable (dogs in the tree, cats on the ground) or out of equilibrium (cats and dogs on the ground). Now, we clear these animals out and release a few dozen white cats and black cats into the yard. The probability that a cat will take to the tree or the ground is independent of the color of its coat, the energy of interaction is low, and the most likely arrangement is one of maximum entropy where the proportion of white cats and black cats is the same at each site. Our black cats and white cats are isotopes with

Elemental
fractionation

Isotopic
homogenization

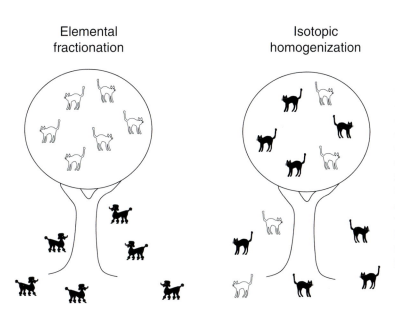

Figure 3.1: Left: Cats and dogs interact vigorously affecting site occupation (tree or yard). Just like two elements with different chemical properties, they arrange themselves so as to achieve the most stable configuration. Right: White cats and black cats have very similar properties and like isotopes of the same element are arranged randomly among the available sites. The most likely arrangement is an identical proportion of isotopes in each site.

very similar properties that share the sites evenly whatever their energy level.

If elements and their isotopes are allowed to move easily between sites in crystals, liquids, and gases, either because the liquid states enable effective mixing or because thermal diffusion allows atoms to move rapidly, the elements with variable properties will arrange themselves in accessible sites so as to minimize the total energy of the system. On the other hand, isotopic exchanges of a single element between phases contribute little to the energy balance of the system and such isotopes will be evenly distributed so as to maximize the entropy of the system. We need not go into the existence of isotopic fractionation, a phenomenon which is of little importance in heavy elements. For the needs of geochronology, both natural mass fractionation and instrumental mass bias are purely and simply eliminated by internal normalization against some arbitrary reference ratio (see Appendix F). As carbonates precipitate out from seawater, the $^{87}Sr/^{86}Sr$ ratio is exactly the same in the calcite as in the seawater from which it precipitates; as the mantle melts, $^{143}Nd/^{144}Nd$ is the same in the molten liquid as in the residue. There are, admittedly, processes where melting occurs in disequilibrium, but even if thermodynamic fractionation persisted at such high temperatures it would be corrected out by the internal standardization procedure. In the rest of the discussion, we will therefore ignore mass-dependent fractionation.

In (3.2), let us divide P by the number P' of atoms of a stable isotope of the same element as the radioactive nuclide. As the system is closed, the number of stable nuclides P' remains constant, which we denote

$P' = P'_0$. This gives:

$$\left(\frac{P}{P'}\right)_t = \left(\frac{P}{P'}\right)_0 e^{-\lambda t} \tag{3.4}$$

The additional condition required to make the decay equation a chronometer is no longer to assume P_0 but rather to determine the isotope ratio $(P/P')_0$ when the system formed, which is already a far less restrictive condition. This method is employed for many short-lived nuclides (^{14}C, ^{10}Be, ^{210}Pb) created by solar or galactic radiation interacting with the atmosphere or rocks (cosmogenic nuclides). In the case of the carbon-14 clock, P refers to the radioactive ^{14}C isotope, while P' is the most abundant isotope ^{12}C of carbon, and a hypothesis is made about the isotopic abundance of ^{14}C in the upper atmosphere. The principle of standardization to a stable isotope is also utilized for radioactive nuclides derived from the decay of uranium isotopes (^{234}U, ^{230}Th, ^{231}Pa), but the equations are then a little more complex.

Equation (3.3) may also be divided by the number D' of atoms of a stable isotope of the same element as the radiogenic nuclide. As the system is closed, the number of stable nuclides remains constant and $D' = D'_0$. This yields:

$$\left(\frac{D}{D'}\right)_t = \left(\frac{D}{D'}\right)_0 + \left(\frac{P}{D'}\right)_t (e^{\lambda t} - 1) \tag{3.5}$$

Equation 3.5 is known as the isochron equation: in a plot of $x = P/D'$ and $y = D/D'$, a set of sub-systems of the same age T and the same initial isotope ratio $(D/D')_0$ will lie on a straight line of slope $e^{\lambda t} - 1$. The P/D' ratio is usually referred to, somewhat improperly, as the parent/daughter ratio. Thus for the system ^{87}Rb–^{87}Sr, P relates to ^{87}Rb, D to ^{87}Sr, and D' to ^{86}Sr, and we can write:

$$\left(\frac{^{87}Sr}{^{86}Sr}\right)_t = \left(\frac{^{87}Sr}{^{86}Sr}\right)_0 + \left(\frac{^{87}Rb}{^{86}Sr}\right)_t (e^{\lambda_{87Rb} t} - 1) \tag{3.6}$$

This expression has the familiar form of the equation of a straight line $y = y_0 + mx$, where $x = (^{87}Rb/^{86}Sr)_t$, $y = (^{87}Sr/^{86}Sr)_t$, with intercept $y_0 = (^{87}Sr/^{86}Sr)_0$ and slope $m = e^{\lambda_{87Rb} t} - 1$.

Figure 3.2 shows two important examples of isochron diagrams. The existence of an alignment of points in the isochron diagram, in which relations between the quantities plotted change with time, cannot be coincidental with the presence of the observer: if the samples form an alignment at the present time, an alignment must have existed at each time since their formation. Since the D/D' ratio of any sample devoid of parent nuclide is unchanged, alignments simply rotate with time around the intercept.

Let us now take a quick guided tour of dating methods, beginning with those based on the measurement of radioactive nuclides; then the

Table 3.1: *Decay constants of the major radioactive systems: the daughter nuclide is shown when it is used for dating*

System	λ (y^{-1})	System	λ (y^{-1})	System	λ (y^{-1})
^{138}La–^{138}Ce	2.24×10^{-12}	^{40}K–^{40}Ca	4.96×10^{-10}	^{26}Al	9.80×10^{-7}
^{147}Sm–^{143}Nd	6.54×10^{-12}	^{235}U–^{207}Pb	9.85×10^{-10}	^{36}Cl	2.30×10^{-6}
^{87}Rb–^{87}Sr	1.42×10^{-11}	^{146}Sm–^{142}Nd	6.73×10^{-9}	^{230}Th	9.20×10^{-6}
^{187}Re–^{187}Os	1.64×10^{-11}	^{244}Pu	8.66×10^{-9}	^{234}U	2.83×10^{-6}
^{176}Lu–^{176}Hf	1.93×10^{-11}	^{182}Hf–^{182}W	7.7×10^{-8}	^{231}Pa	2.11×10^{-5}
^{232}Th–^{208}Pb	4.95×10^{-11}	^{129}I–^{129}Xe	4.30×10^{-8}	^{14}C	1.21×10^{-4}
^{40}K–^{40}Ar	5.81×10^{-11}	^{53}Mn–^{53}Cr	1.87×10^{-8}	^{226}Ra	4.33×10^{-4}
^{238}U–^{206}Pb	1.55×10^{-10}	^{10}Be	4.62×10^{-7}	^{210}Pb	3.11×10^{-2}

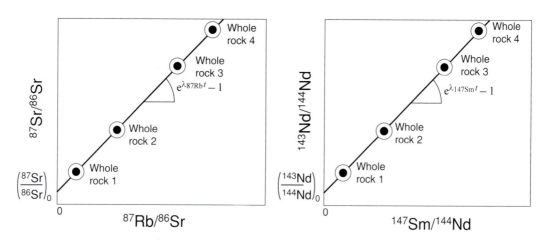

Figure 3.2: Two examples of isochrons: the ^{87}Rb $\rightarrow$ ^{87}Sr system (left) and the ^{147}Sm $\rightarrow$ ^{143}Nd system (right). Samples of the same age, and for which the isotopic composition of the daughter element can be considered identical, will lie on a straight line, the slope of which gives the age of formation. The intercept of the straight line gives the isotopic ratio of the daughter element at that particular moment.

systems with high parent/daughter ratios, in which the initial quantities of radiogenic nuclides can be neglected; and, finally, systems with low parent/daughter ratios, where the isochron method is applicable. In general, a clock can be applied to samples whose age does not exceed five times the radioactive period. Table 3.1 shows that the clocks spread over a wide range but certain age-ranges are not well covered, especially that at around one million years.

Note that physical time elapsing in the real world is normally given in seconds (s), which is not a very helpful unit in the Earth sciences, while geological ages, through which we go back through time, are noted in anni (a), from the Latin *annus*. Derived units ky and ka (thousand years), My and Ma (million years), Gy and Ga (billion years) apply to time and age, respectively. Appendix E shows the division of geological time into absolute ages.

3.1 Dating by radioactive nuclides

This group of methods relates essentially to nuclides produced by cosmic radiation, but we will see that the approach can be generalized to the descendants of uranium and thorium with methods based on the surpluses of these nuclides. Here we make an assumption about the initial isotopic composition of the element to which the radioactive nuclide belongs.

3.1.1 Carbon-14

This method of dating, which is certainly the most familiar to the general public, is not the oldest historically. However, it has revolutionized archeology and earned its inventor, Libby, the Nobel Prize for Chemistry in 1960. The Earth is subjected to bombardment from high-energy galactic cosmic rays, mostly protons and α particles, which react with the Earth's atmosphere. The interaction of these particles with nitrogen and oxygen produces secondary particles, mostly neutrons. An important reaction is that of these cosmic neutrons with nitrogen, which produces radioactive carbon-14 and a proton:

$$^{14}N + n \rightarrow {}^{14}C + p \tag{3.7}$$

The ^{14}C atom decays to ^{14}N by β^- emission with a decay constant λ_{14C} of $1.2 \times 10^{-4}\,y^{-1}$. Before disappearing, it mixes very quickly with the stable isotopes of carbon, the most important of which is ^{12}C. Equation 3.4 can now be written:

$$\left(\frac{^{14}C}{^{12}C}\right)_t = \left(\frac{^{14}C}{^{12}C}\right)_0 e^{-\lambda_{14C}t} \tag{3.8}$$

Plants exchange their carbon with the atmosphere, with which they are in isotopic equilibrium until they die. It can be seen that if the ratio $(^{14}C/^{12}C)_0$ in the atmosphere is constant and known, measurement of the $^{14}C/^{12}C$ ratio in wood or a fossil carbonate will date the death of the organism.

This approach is complicated by several effects. First, ever since the 19th century, the burning of coal and oil has released a large quantity of "dead" carbon devoid of ^{14}C into the atmosphere, thus complicating estimates of the $(^{14}C/^{12}C)_0$ ratio. Moreover, above-ground nuclear explosions until the mid 70s have contaminated the atmosphere with artificial

^{14}C, some of which has invaded the surface of the oceans. Finally, the variation in solar activity modulates galactic cosmic radiation received by the Earth and therefore changes the rate of production of ^{14}C in the atmosphere over very long periods. To overcome these difficulties, the ^{14}C scale has been calibrated for more recent ages by dendrochronology – a method based on counting the growth rings of very old trees – Californian bristle-cone pines or German oaks. For older ages, calibration is achieved by comparison with the thorium-230 method on corals.

There are various applications for the ^{14}C method; it can provide dates as old as 40 000 years ago. Measurement methods using a linear accelerator have pushed this limit a little further back, but above all they have reduced the quantities of material required for analyses to be conducted. The ^{14}C method has been used very successfully in archeology and Quaternary geology. It also has applications in dating ground water sources and, as we will see later, deep water of the oceans.

3.1.2 Beryllium-10

The radioactive nuclide ^{10}Be is produced by the effect of cosmic rays on atmospheric ^{14}N and ^{16}O. Metallic Be is rapidly oxidized as BeO, scavenged by atmospheric particles, and finally incorporated into soil and sediments by rain water and runoff. The ^{10}Be decay constant is 4.62×10^{-7} y^{-1} and it is customary to normalize its abundance to that of the stable isotope ^{9}Be. Beryllium is an element similar to aluminum and is found in clay and soils. Beryllium-10 dating is much used in oceanography for measuring sedimentation rates or manganese nodule growth rates. For samples taken at depth z in a sediment core in which the sedimentation rate can be assumed to have remained constant, the equivalent to (3.8) can be written as:

$$\left(\frac{^{10}\text{Be}}{^{9}\text{Be}}\right)_t = \left(\frac{^{10}\text{Be}}{^{9}\text{Be}}\right)_0 \exp\left(-\lambda_{10\text{Be}}\frac{z}{v}\right) \tag{3.9}$$

where v is the sedimentation rate. On the condition that the ^{10}Be/^{9}Be ratio at the time of deposition does not vary locally over the time interval sampled by the core, the ^{10}Be/^{9}Be ratio measured in different samples should decrease exponentially with their age. The logarithm of the ^{10}Be/^{9}Be ratio should therefore vary linearly with depth z (Fig. 3.3) in the core. The slope $-\lambda_{10\text{Be}}/v$ is therefore a measure of sedimentation rate.

This method can also be employed for measuring erosion rates. Neutrons produced by interaction of cosmic radiation with the atmosphere are stopped by the ground, where they cause spallation reactions, i.e. fragmentation of the atoms of the material. The production of ^{10}Be from silicon atoms in quartz is of particular interest. The number

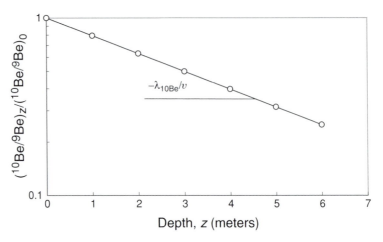

Figure 3.3: Measurement of sedimentation rate v by variation in relative abundance of the cosmogenic isotope ^{10}Be with depth. The isotopic ratio is standardized to its surface value (0). The scale is semi-logarithmic.

of ^{10}Be atoms produced in the ground per unit volume varies with depth z beneath the surface as $R_0\, e^{-z/l}$, where R_0 is the surface production and l is the attenuation distance or mean penetration distance of cosmic particles into the rock. Over time t, thickness $dz = vdt$ of the rock has been eroded, where v is the erosion rate. The number dN of ^{10}Be atoms transiting between depth z and depth $z + dz$ is $-(vN(z + dz) - vN(z))$ per unit time, since the erosion rate is directed toward negative depths. Let us assume that the erosion rate is constant and that the ^{10}Be production rate remains constant over the time interval of interest. At equilibrium we can write that removal by erosion and radioactivity balances production by cosmic radiation:

$$-\left[vN(z + dz) - vN(z)\right] + \left[R_0 e^{-z/l} - \lambda N(z)\right] dz = 0 \qquad (3.10)$$

and a little mathematics gives the expression:

$$N(z) = \frac{R_0}{\lambda + v/l} e^{-z/l} \qquad (3.11)$$

It can be seen, then, that if we know the rate of production R_0 and the attenuation distance l, the abundance $N(z = 0)$ of ^{10}Be at the surface provides a direct estimate of the rate v of erosion.

3.1.3 The thorium-230 excess method

This is a somewhat different case, as this nuclide, with a radioactive half-life of 75 000 years, is not created by radiation but by another parent nuclide whose half-life is long enough for its rate of production to be considered constant on time scales of less than one million years. It is one of the examples of clocks based on a chain of radioactive decay (Fig. 3.4) in which nuclides decay from one to another by α or β^-

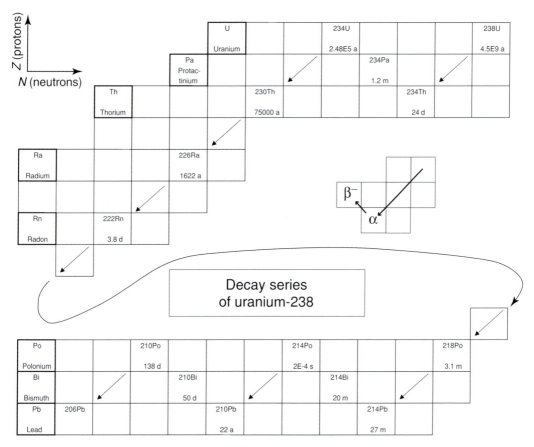

Figure 3.4: Decay series of ^{238}U. Chain decay of ^{238}U. The decay period of unstable nuclides is shown. The number of neutrons is given on the x-axis and the atomic number on the y-axis. Jumps induced by α and β^- decay are shown in the inset box. Units: year (a), day (d), second (s).

radioactive processes. The vast majority of intermediate nuclides have half-lives too short to act as useful clocks. There are only four radioactive chains. These chains, which have a long-lived, heavy nuclide as their parent, are those of ^{232}Th, ^{235}U, ^{238}U, and ^{237}Np. The first three end with three isotopes of lead, ^{208}Pb, ^{207}Pb, ^{206}Pb, and we will see that their relative abundances in modern lead are utilized as clocks. The fourth chain is that of neptunium-237, an extinct radiogenic nuclide ending with the single isotope bismuth-209. It is not detectable in natural products.

Uranium-238 decays to ^{234}Th, then to ^{234}U and, finally, to ^{230}Th, which is itself radioactive. The two intermediate nuclides, ^{234}Th and ^{234}U, are so short-lived that we can ignore them here. The change in the number of ^{230}Th nuclides can therefore be written as the difference between production by the parent ^{238}U and radioactive decay:

$$\frac{d^{230}Th}{dt} = -\lambda_{230Th} \, ^{230}Th + \lambda_{238U} \, ^{238}U \tag{3.12}$$

After a few hundreds of thousand years, the number of ^{230}Th nuclides reaches steady state and the two terms of the right-hand side cancel out. This state, where all activity levels are equal for all the nuclides in the chain, is known as secular equilibrium. The activity $\left[^{238}U\right]$ remains essentially constant during the time it takes to establish secular equilibrium (roughly $1/\lambda_{230Th}$, or several 100 000 years) and variation d $\left[^{238}U\right]/dt$ with time is therefore zero. Allowing for this property and multiplying (3.12) by λ_{230Th}, we obtain:

$$\frac{d\left\{\left[^{230}Th\right] - \left[^{238}U\right]\right\}}{dt} = -\lambda_{230Th} \left\{\left[^{230}Th\right] - \left[^{238}U\right]\right\} \tag{3.13}$$

in which the square brackets represent activity. The difference $\left[^{230}Th\right] - \left[^{238}U\right]$ is known as ^{230}Th excess (the term excess is taken by reference to the amount present at secular equilibrium), and is written $\left[^{230}Th\right]_{ex}$. Equation (3.13) integrates as:

$$\left[^{230}Th\right]_{ex} = \left[^{230}Th\right]_{ex,0} e^{\lambda_{230Th}t} \tag{3.14}$$

where the subscript 0 denotes the initial time. If, as with ^{10}Be, we wish to measure a sedimentation rate, time t is replaced by the ratio between depth z and rate of sedimentation v, and the logarithm of $\left[^{230}Th\right]_{ex}$ is plotted as a function of z for several samples collected at different depths from a single core: the slope of the alignment indicates the $-\lambda_{230Th}/v$ ratio, and so v can be determined from it.

These concepts apply to excesses of several nuclides descending from ^{238}U, such as ^{234}Th, ^{234}U, ^{210}Pb, and from ^{235}U, such as ^{231}Pa. These methods are often said to be based on "uranium disequilibrium series," to indicate that the chronometric information lies with the deviation of a given parent/daughter pair from secular equilibrium. There are many fields of application, from oceanography to dating of Quaternary lacustrine or marine sediments. Lead-210 (^{210}Pb), with a half-life $t^{1/2} = 23.3$ years, notably forms in the atmosphere and in rainwater by decay of radon-222 (the offspring of ^{238}U, see Fig. 3.4), a gaseous nuclide given off constantly by the ocean, mantle, and crust. It is used for determining the rates of accumulation of ice or of sedimentation during the last century. It is also very helpful for studying pollution processes.

3.2 Systems with high parent/daughter ratios

There are many important clocks for which the quantities of daughter isotopes at time $t = 0$ can be ignored. The potassium–argon method and uranium–lead method on zircon are the best known. Also worth mentioning in this category are the many clocks based on the descendants of uranium, such as ^{238}U–^{230}Th, which have been amazingly successful at dating corals and cave deposits.

3.2.1 The potassium–argon method

This is the workhorse of geochronology. The method relies on the potassium-40 nucleus capturing a K-shell electron so that ^{40}K + e$^-$ → ^{40}Ar. The radioactive constant or probability of decay per unit time for this process is $\lambda_\varepsilon = 5.81 \times 10^{-11}$ y^{-1}. Argon-40 also decays by an ordinary β^- process into ^{40}Ca (dual decay), for which the radioactive constant is $\lambda_\beta = 4.96 \times 10^{-10}$ y^{-1}. The proportion of ^{40}K atoms taking the ^{40}Ar pathway is equal to the relative probability $\lambda_\varepsilon/(\lambda_\varepsilon + \lambda_\beta)$ or 10.5%. Equation (3.3) then becomes:

$$^{40}\mathrm{Ar}_t = \,^{40}\mathrm{Ar}_0 + \frac{\lambda_\varepsilon}{\lambda_\varepsilon + \lambda_\beta} \,^{40}\mathrm{K}_t \left[e^{(\lambda_\varepsilon + \lambda_\beta)t} - 1 \right] \qquad (3.15)$$

with the sum of probabilities of decay, by one or other pathway, in the exponential, and the proportion of daughter nuclides that are atoms of ^{40}Ar in the factor term. The principle behind the method is that the ^{40}Ar$_0$ term can be neglected relative to the second term. Although argon is present in notable quantities in the atmosphere (1%) and in the interstitial gases of rocks, this inert gas is not very soluble at ambient pressure in melts and minerals since it forms only weak van der Waals' type bonds with mineral ions. At higher pressure, however, substantial amounts of argon may remain trapped in submarine glasses or metamorphic minerals (excess argon).

For analysis, argon is extracted from rocks by heating and melting in ultra-vacuum lines and analyzed with a mass spectrometer. The ultra-vacuum is necessary to prevent atmospheric contamination. Such contamination inevitably occurs as the minerals adsorb small quantities of atmospheric gases on their surfaces or in grain fractures. To obviate this, we use the fact that atmospheric argon has several isotopes and that the ^{40}Ar/^{36}Ar ratio is 296 in the atmosphere. We just subtract 296 times the quantity of ^{36}Ar measured, from the ^{40}Ar measured in the sample, to obtain the radiogenic argon (Fig. 3.5). This can only be done with precision when radiogenic ^{40}Ar is not dominated by that originating in the atmosphere: measuring young ages for rocks with a low potassium content is therefore a technical feat of skill. It is thanks to the potassium–argon method that the scale of magnetic reversals can be calibrated, as it is the only method for dating the lava from which the paleomagnetic measurements are made. It will be seen when studying the

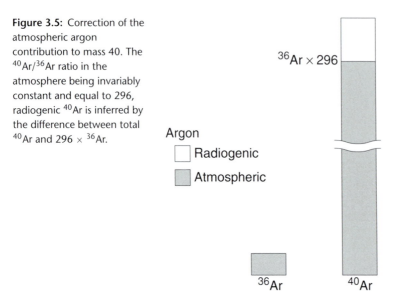

Figure 3.5: Correction of the atmospheric argon contribution to mass 40. The $^{40}Ar/^{36}Ar$ ratio in the atmosphere being invariably constant and equal to 296, radiogenic ^{40}Ar is inferred by the difference between total ^{40}Ar and $296 \times {}^{36}Ar$.

thermal history of rocks in the chapter on diffusion, that it also provides a measure of post-orogenic cooling rates and rates of exhumation of mountain ranges. Applications in other fields find it difficult to compete with other methods, notably U–Pb on zircons.

An important and widely used variant of the standard ^{40}K–^{40}Ar method involves irradiating the sample with rapid neutrons. It is not ^{40}K, but the ^{39}Ar produced when the neutrons react with the ^{39}K isotope that is measured. Accordingly, this is termed the ^{39}Ar–^{40}Ar method, which replaces the separate measurement of two isotopes of two different elements, ^{40}K and ^{40}Ar, with a precise measurement of the $^{40}Ar/^{39}Ar$ ratio of two isotopes of argon. A control sample of known age (monitor), irradiated at the same time, provides a measure of the yield of the reaction. This method, combined with progressive extraction of argon from the irradiated sample, has several advantages, described in more specialized texts, and include the identification of possible argon losses after the formation of minerals.

3.2.2 Dating zircons by the uranium–lead method

This is the Ferrari of geochronology for long geological time scales; it is difficult to implement, but precise, and is resistant to disturbances occurring after closure of the system. In recent years, its development has considerably benefited from the improved cleanliness of chemical extractions and *in situ* methods of analysis (secondary ion mass spectrometry). The advantage of the method lies in the radioactive (^{238}U and

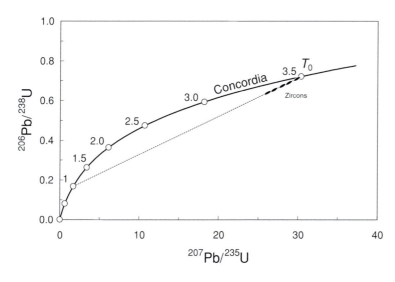

Figure 3.6: The concordia curve is the locus of points for which the x- and y-axes yield identical ages for both the ^{235}U–^{207}Pb and ^{238}U–^{206}Pb methods. The numbers 1, 2.0, 2.5, etc., on the curve correspond to geological ages of 1, 2.0, 2.5, etc., billions of years. The zircons extracted from a single sample, and shown here by the black ellipses, indicate an age T_0 of crystallization of 3.5 Ga, with 1.0 Ga old overgrowth. The dashed lined is a mixing line between the two zircon generations.

^{235}U) and radiogenic nuclides (^{206}Pb and ^{207}Pb) being isotopes of the same element: uranium for one and lead for the other. In the absence of any initial radiogenic lead, (3.5) applied to the systems ^{238}U–^{206}Pb ($\lambda_{238U} = 0.155\ 125 \times 10^{-9}$ y^{-1}) and ^{235}U–^{207}Pb ($\lambda_{235U} = 0.984\ 85 \times 10^{-9}$ y^{-1}) gives:

$$\left(\frac{^{206}Pb}{^{238}U}\right)_t = e^{\lambda_{238U}t} - 1$$

$$\left(\frac{^{207}Pb}{^{235}U}\right)_t = e^{\lambda_{235U}t} - 1 \tag{3.16}$$

This double clock is routinely applied to the radiogenic lead and uranium of an accessory, but common, zirconium silicate of granite and metamorphic rocks, zircon ($ZrSiO_4$). Uranium U^{4+} substitutes in large quantities for Zr^{4+}; but Pb^{2+}, which is of very different ionic radius and charge from Zr^{4+}, is essentially excluded at equilibrium. As with atmospheric argon, there may nonetheless be contamination by lead at mineral surfaces or in grain fractures. Because of the presence of tetra-ethyl lead, used as an anti-knocking agent in fuel, man-made pollution may also be significant. A very similar technique to that described for argon is employed, involving subtraction from the total lead content the contaminated lead, whose isotopic composition is relatively well known by using a stable isotope ^{204}Pb. *In situ* isotopic analysis using modern ion probes also allows the zones for analysis to be selected so that contamination is almost completely eliminated.

The pair of equations in (3.16) defines the locus of points for which the ages indicated by both methods concur, the locus being traditionally called concordia. This concordia flattens out toward older ages (Fig. 3.6), as ^{235}U decays much more rapidly than ^{238}U: natural uranium today contains only 0.7% ^{235}U compared with 8% ^{235}U three billion years ago. Although methods were developed to attempt to correct the effect of disagreement related to losses of lead after closure of the system, they are now of little value because of the improvement in techniques, particularly the mechanical abrasion of zircons, which allow us to concentrate only on their unaltered parts. The ratio of the x-axis to the y-axis is proportional to the isotope ratio ^{206}Pb/^{207}Pb:

$$\frac{\left(^{206}\text{Pb}/^{238}\text{U}\right)_t}{\left(^{207}\text{Pb}/^{235}\text{U}\right)_t} = \frac{1}{\left(^{207}\text{Pb}/^{206}\text{Pb}\right)_t} \times \frac{1}{\left(^{238}\text{U}/^{235}\text{U}\right)_t} \tag{3.17}$$

The second term on the right-hand side is constant and equal to 1/137.88. The denominators of the ratios plotted on the x- and y-axes are therefore proportional and their ratio ^{238}U/^{235}U is constant. It follows from the discussion of ratio behavior during mixing, presented earlier (Fig. 2.4), that mixtures of zircons or overgrowth will be reflected by alignments in the concordia plot: the intercepts of these alignments with the concordia will therefore a priori yield interesting ages.

3.3 The isochron method

When minerals and rocks form, they already contain some of the radiogenic isotopes used for dating. The daughter nuclide may be present in large, yet unknown, concentrations. The isochron method was devised to provide an age, even when the amount of radiogenic isotope initially present in the system is not negligible with respect to that produced by radiogenic decay after its formation. The key assumption is that the initial isotopic composition of the element to which the radiogenic nuclide belongs is unknown, but constant, in all the samples analyzed. Isotopic homogenization is assumed to be complete at $t = 0$, which may be the case where minerals crystallize from a magma or from seawater within a time that can be considered as very short compared to the age of the rocks. A large number of geochronological systems are used in this way: ^{87}Rb–^{87}Sr, ^{147}Sm–^{143}Nd, and ^{187}Re–^{187}Os are examples (note that chronological systems are denoted by hyphenating the parent and daughter isotopes in the order P-D). In addition to the chronological aspect, the variations in the isotopic abundances of radiogenic isotopes are useful in studying many geological processes. Let us take as an example the ^{147}Sm–^{143}Nd system, for which the stable reference isotope is usually ^{144}Nd, and in (3.5) replace P with ^{147}Sm, D with ^{143}Nd, and D' with ^{144}Nd. For a closed system, the isochron equation becomes:

$$\left(\frac{^{143}\text{Nd}}{^{144}\text{Nd}}\right)_t = \left(\frac{^{143}\text{Nd}}{^{144}\text{Nd}}\right)_0 + \left(\frac{^{147}\text{Sm}}{^{144}\text{Nd}}\right)_t \left(e^{\lambda_{147\text{Sm}}t} - 1\right) \tag{3.18}$$

Samples, for example basalts extracted from a single source in the mantle, formed at t with the same $(^{143}\text{Nd}/^{144}\text{Nd})_0$ ratio but with different $^{147}\text{Sm}/^{144}\text{Nd}$ ratios, presumably by different degrees of melting, will form an alignment known as the isochron (Fig. 3.2), whose slope gives the age of formation. It is commonly held that, for the isochron to be valid, the rocks must be "co-genetic" and the initial isotopic homogeneity must be perfect. However, as a result of improved analytical techniques and instrumentation, there is always an achievable level of precision for which the assumption of initial isotopic homogeneity breaks down. A less conservative, but more realistic, statement is that the initial isotopic variability must remain negligible compared with that related to the accumulation of radiogenic nuclides: this explains why isochrons of old age, such as are obtained from meteorites or lunar rocks, "look" better than those from recent granites, or why isochrons of reasonable age can be obtained from samples that are not necessarily taken from a homogeneous medium, such as sedimentary sample suites.

Let us look at the problem of how the isochron withstands subsequent perturbations of the closed-system regime. Let us imagine, for example, a series of basalts formed from a single mantle source in Archean time, some 2.7 billion years ago, and subjected to intense thermal perturbation when, caught up in the formation of a mountain belt some 600 million years ago, they were transported to great depth and raised to temperatures of the order of 600–800 °C. Such metamorphism will create new minerals in basaltic rocks, probably an assemblage of pyroxene and garnet known as eclogite. Suppose that a mean distance d can be defined over which the elements Sm and Nd migrate in the course of metamorphic recrystallization, ranging probably from a few millimeters to a few centimeters (Fig. 3.7). If our sample is very much smaller than d, we can presume that the exchanges of Sm and Nd are complete and that the minerals, normally of very small size, will all have adopted the ambient Nd isotopic composition, i.e. that of the rock at this location. If the sample size is much greater than d, only the outer part over a depth d will have engaged in exchange with the exterior. The interior will have been disturbed, but the exchanges will have remained confined within the sample. If this open fringe represents only a small volume compared with the bulk of the sample, it will be considered that the sample has not been affected by the disturbance.

This pattern of exchange suggests the sampling strategy (Fig. 3.8). To determine the age of any metamorphic perturbation, the minerals of a rock are separated, here garnet and pyroxene, whose small size ensures that they will have been returned to isotopic equilibrium during metamorphism. The age indicated by what is termed the internal

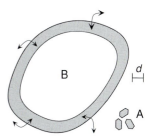

Figure 3.7: Closed system and open system. If we imagine that during a metamorphic perturbation an element has moved an average distance d, small systems (minerals A) will be "opened" much more intensely than large ones (large sample B), where only the outer skin will have exchanged with the ambient medium. The shaded area represents the material involved in exchange outside the system.

Figure 3.8: During a metamorphic event, large samples (whole rock) will remain virtually closed (see Fig. 3.7) and the "external" isochron will record the age T_0 of rock formation. However, the minerals of rock 2 (pyroxene, garnet) will have exchanged their isotopes and their ^{143}Nd/^{144}Nd isotope ratios will be in equilibrium with that of the mean ambient value, i.e. that of the rock. The "internal" isochron therefore gives the age T_1 of the disturbance.

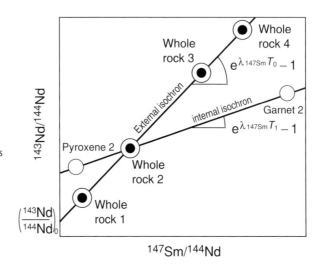

isochron (or mineral) will be that of the metamorphic event. Conversely, to obtain the age of formation of the rock, the largest possible samples must be taken so as to minimize the effect of exchanges with the exterior. The whole rock isochron will be obtained from small fractions of large samples that have been thoroughly ground and mixed.

There are many methods of dating using isochrons. Historically, ^{87}Rb–^{87}Sr chronology has yielded many ages for the emplacement of granites, whereas the dating of white mica (muscovite) or black mica (biotite), whose high ^{87}Rb/^{86}Sr ratio ensures good age precision, have dominated metamorphic geochronology alongside ^{40}K–^{40}Ar ages. The mobility of rubidium, an alkaline element, and of strontium, an alkaline-earth element, in metamorphic and hydrothermal fluids unfortunately often disturbed this chronometer. ^{147}Sm–^{143}Nd chronology is valuable for dating old basaltic rocks. Samarium and neodymium, two of the rare-earths, are much less mobile, but the method has also been known to go awry. Dating high-temperature and high-pressure metamorphism, especially with pyroxene–garnet internal isochrons, has yielded good results by this method, as has, more recently, the ^{176}Lu–^{176}Hf system. Dating of basaltic rocks and peridotites by the ^{187}Re–^{187}Os method, two elements of the platinum group, has also proved to be very successful.

A first special application of the isochron method is that of the ^{230}Th dating method, which is especially useful for recent volcanic rocks. If we re-write (3.14) by allowing for the definition of excess ^{230}Th and if we divide both sides by the activity of the thorium-232 isotope (which decays slowly enough for this activity to be considered constant),

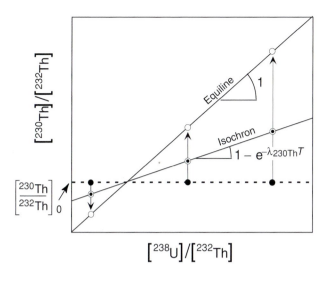

Figure 3.9: The U–Th isochron. Samples derived from the same sedimentary or magmatic reservoir initially have the same $[^{230}\text{Th}/^{232}\text{Th}]$ isotope ratio but different U/Th ratios. They line up first on a horizontal line and then migrate progressively toward the equiline, which is the stable state of secular equilibrium. Square brackets indicate activity.

we obtain:

$$\left(\frac{[^{230}\text{Th}]}{[^{232}\text{Th}]}\right)_t = \left(\frac{[^{230}\text{Th}]}{[^{232}\text{Th}]}\right)_0 e^{\lambda_{230\text{Th}}t} + \left(\frac{[^{238}\text{U}]}{[^{232}\text{Th}]}\right)_t \left(1 - e^{\lambda_{230\text{Th}}t}\right) \qquad (3.19)$$

It can be seen from an isochron plot $x = [^{238}\text{U}] / [^{232}\text{Th}]$, $y = [^{230}\text{Th}] / [^{232}\text{Th}]$ (Fig. 3.9) that samples formed at $t = 0$ with the same $[^{230}\text{Th}] / [^{232}\text{Th}]_0$ ratio will lie on a horizontal line. With time the alignment pivots around the point of intersection with the line $y = x$, known as the equiline, until its slope becomes unity: at this point, the system is in secular equilibrium $([^{230}\text{Th}] = [^{238}\text{U}]$, see above). The slope of the alignment gives an age, at least as long as we are far from equilibrium. This method is often applied to minerals extracted from lavas to date their crystallization. It can only be applied for ages of less than 350 000 years.

A second instance is that of the lead–lead method. By utilizing the stable lead isotope of mass 204, the two equations of the isochron can be written:

$$\left(\frac{^{206}\text{Pb}}{^{204}\text{Pb}}\right)_t - \left(\frac{^{206}\text{Pb}}{^{204}\text{Pb}}\right)_0 = \left(\frac{^{238}\text{U}}{^{204}\text{Pb}}\right)_t \left(e^{\lambda_{238\text{U}}t} - 1\right) \qquad (3.20)$$

$$\left(\frac{^{207}\text{Pb}}{^{204}\text{Pb}}\right)_t - \left(\frac{^{207}\text{Pb}}{^{204}\text{Pb}}\right)_0 = \left(\frac{^{235}\text{U}}{^{204}\text{Pb}}\right)_t \left(e^{\lambda_{235\text{U}}t} - 1\right) \qquad (3.21)$$

By dividing (3.21) by (3.20), we obtain:

$$\frac{\left(^{207}\text{Pb}/^{204}\text{Pb}\right)_t - \left(^{207}\text{Pb}/^{204}\text{Pb}\right)_0}{\left(^{206}\text{Pb}/^{204}\text{Pb}\right)_t - \left(^{206}\text{Pb}/^{204}\text{Pb}\right)_0} = \left(\frac{^{235}\text{U}}{^{238}\text{U}}\right)_t \frac{e^{\lambda_{235\text{U}}t} - 1}{e^{\lambda_{238\text{U}}t} - 1} \qquad (3.22)$$

Figure 3.10: ^{206}Pb/^{204}Pb and
^{207}Pb/^{204}Pb ratios of samples
formed with the same initial
isotopic composition of lead
but different U/Pb ratios
evolve along growth curves so
that the isotopic ratios remain
on an isochron with a slope
indicative of the age of
formation.

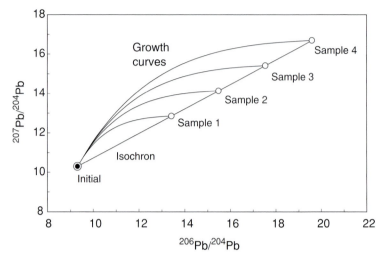

Because the present ratio ^{235}U/^{238}U is a constant equal to 1/137.88, (3.22) can be recast as $(y - y_0)/(x - x_0) = m$. In a plot $x = \left(^{206}\text{Pb}/^{204}\text{Pb}\right)_t$, $y = \left(^{207}\text{Pb}/^{204}\text{Pb}\right)_t$ (Fig. 3.10), (3.22) describes an isochron straight line going through the point of coordinates (x_0, y_0) and with slope m. The advantage of this method is that it requires only the isotope ratios of lead to be determined and not the concentrations, in particular that of uranium which is commonly strongly affected by water circulation in the water table and by weathering. For this reason, this method was commonly used for dating all sorts of sedimentary and magmatic rocks until superseded by zircon geochronology. It will be seen that this was the first method ever to yield the age of the Solar System and it is still widely used to date meteorites and planetary samples.

3.4 Radiogenic tracers

The initial isotopic ratios $(D/D')_0$ obtained as intercepts of isochrons were once a by-product of geochronology. These ratios, which describe the value of ^{87}Sr/^{86}Sr, ^{143}Nd/^{144}Nd, ^{176}Hf/^{177}Hf, etc., at the time the rocks formed, have become, together with the isotopic ratios measured in modern samples, a prime source of geochemical information about parent/daughter ratios in various geological systems. If (3.18) is considered not as an isochron equation but as expressing a change in the ^{143}Nd/^{144}Nd ratio versus time (Fig. 3.11), we have a means of evaluating the mean Sm/Nd ratio of the system. Let us try to recast this equation into a different form using (3.1), in which the parent nuclide P is ^{147}Sm. In our usual notation, the daughter nuclide D is ^{143}Nd and the stable nuclide D' is ^{144}Nd. Using the condition of nuclide conservation during

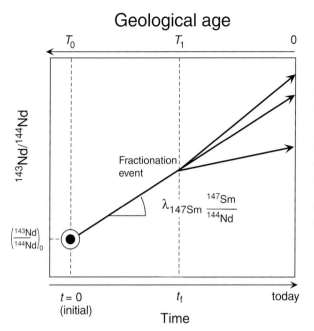

Figure 3.11: Evolution of the ^{143}Nd/^{144}Nd isotopic ratio by radiogenic decay of ^{147}Sm. This ratio changes almost linearly with time t and the slope of the evolution curve indicates the ^{147}Sm/^{144}Nd, and therefore the Sm/Nd ratio, of the system. Magmatic, sedimentary, or metamorphic events, here represented as occurring at T_1, may fractionate the Sm/Nd ratio of various sub-systems.

decay and dividing the equation by ^{144}Nd, we obtain:

$$\frac{d}{dt}\left(\frac{^{143}Nd}{^{144}Nd}\right)_t = -\frac{d}{dt}\left(\frac{^{147}Sm}{^{144}Nd}\right)_t = \lambda_{147Sm}\left(\frac{^{147}Sm}{^{144}Nd}\right)_t \qquad (3.23)$$

In other words, the slope of the evolution curve (the so-called growth curve) in Fig. 3.11 is simply $\lambda_{147Sm}(^{147}Sm/^{144}Nd)$, which, given the very slow decay of ^{147}Sm, is almost invariant. If the ^{143}Nd/^{144}Nd ratio is known both at $t = 0$, for example at the origin of the Solar System by utilizing the y-intercept of a meteorite isochron, and today, the mean ^{147}Sm/^{144}Nd ratio prevailing as the system evolved can be deduced:

$$\left(\frac{^{147}Sm}{^{144}Nd}\right)_{average} \approx \frac{1}{\lambda_{147Sm}t}\left[\left(\frac{^{143}Nd}{^{144}Nd}\right)_{today} - \left(\frac{^{143}Nd}{^{144}Nd}\right)_{initial}\right] \qquad (3.24)$$

By dividing by the appropriate isotopic abundance factor (here 0.15/0.24) and by multiplying by the ratio of atomic masses of the two elements (150.4/144.2), this isotope ratio can be converted into a Sm/Nd weight ratio.

Let us compare the ^{143}Nd/^{144}Nd ratio of a basalt from the East Pacific Rise (0.5131) with that of a clay mud collected from the mouth of the Amazon (0.5108). The difference between the ^{87}Sr/^{86}Sr ratios of the two rocks (0.7023 for the basalt and 0.7140 for the granite) is of opposite sign. The same is true of the ^{206}Pb/^{204}Pb ratio (18.0 and 18.8). The basalt source is the upper mantle, whereas the clay is derived from

erosion of South American continental crust. Equation (3.23) can therefore be used to calculate the mean $^{147}Sm/^{144}Nd$ ratio of the upper mantle and that of the continental crust from the beginning of the Earth's
history. Of course, these are virtual ratios, as the crust and mantle have
complex histories that cannot be fully captured by a single parameter, but they give us useful information about the nature of the fractionation processes that have affected these geological units. Meteorite
studies tell us that at the time the Solar System formed 4.56 billion
years ago, the $^{143}Nd/^{144}Nd$ ratio was 0.5067, the $^{87}Sr/^{86}Sr$ ratio 0.6992,
and the $^{206}Pb/^{204}Pb$ ratio 9.3. The $^{147}Sm/^{144}Nd$ ratio of the mantle
in which the neodymium evolved before passing into the basalt was
therefore:

$$\left(\frac{^{147}Sm}{^{144}Nd}\right)^{mantle}_{mean} = \frac{0.5131 - 0.5067}{(0.654 \times 10^{-12}) \times (4.56 \times 10^{9})} = 0.215$$

For the continent that was the source of the clay, this calculation gives
a $^{147}Sm/^{144}Nd$ ratio of 0.138. The $^{87}Rb/^{86}Sr$ ratio of the parent medium
of the two rocks is obtained in a similar way, giving 0.046 for the upper
mantle and 0.22 for the continental crust. For lead, the parent/daughter
ratio varies with time and the linear approximation must be abandoned,
which raises no particular problem. The modern $^{238}U/^{204}Pb$ ratios,
known as μ, are 8.5 and 9.3, respectively.

The Sm/Nd, Rb/Sr, and U/Pb ratios therefore differ between the
continental crust and the upper mantle, so that the more incompatible
element of each pair (Nd, Rb, and U) is more concentrated in the crust
than the corresponding more compatible element (Sm, Sr, and Pb). A
process capable of fractionating this ratio as the continental crust forms
must therefore be imagined, for example, melting followed by selective
extraction of magmas.

By plotting on an isochron diagram the present-day $^{143}Nd/^{144}Nd$ and
$^{147}Sm/^{144}Nd$ ratios of any rock sample of continental crust and those of
the average upper mantle (the MORB source, see Chapter 8), a theoretical
age T_{Nd} can be defined at which this particular piece of continental crust
could have separated from the upper mantle. This age, usually referred
to as the Nd model age of this particular crustal sample, is obtained by
writing (3.18) once for the rock and once for the upper mantle, and by
eliminating the $^{143}Nd/^{144}Nd_0$ ratio between the two expressions:

$$T_{Nd} = \frac{1}{\lambda_{147Sm}} \times \frac{\left(^{143}Nd/^{144}Nd\right)_{rock} - \left(^{143}Nd/^{144}Nd\right)_{mantle}}{\left(^{147}Sm/^{144}Nd\right)_{rock} - \left(^{147}Sm/^{144}Nd\right)_{mantle}} \qquad (3.25)$$

If our Amazon mud has a $^{147}Sm/^{144}Nd$ ratio of 0.120, a value that is
quite representative of the continental crust in general, the model age for

local separation of the crust and the upper mantle is therefore:

$$T_{Nd} = \frac{1}{0.654 \times 10^{-12}} \times \frac{0.5108 - 0.5131}{0.120 - 0.215} = 3.7 \times 10^9 \text{ years}$$

Radiogenic isotopes are most commonly utilized, then, for tracing sources, which we will examine in detail later. The principle is that when rock melts, the parent/daughter ratios are affected in a way that depends on the residual mineral assemblage, while the isotopic ratios remain unaffected. Thus, for mantle melting, Sm/Nd and Lu/Hf ratios are higher in the melt residue than in the liquid, while the contrary is true of Rb/Sr, Re/Os, and U/Pb.

A special case is the production of radiogenic helium ^{4}He, which is achieved essentially by decay of the two natural isotopes of uranium (^{238}U and ^{235}U) and of the only long-lived isotope of thorium (^{232}Th). We have seen earlier that these decay processes are in fact the start of a chain of events. For example, in order to pass from ^{238}U to its distant descendant ^{206}Pb, the initial nuclide must lose $(238-206)/4 = 8$ α particles, which, by capture of the rock matrix electrons torn off during particle expulsion, become that number of ^{4}He atoms. The only stable reference isotope is ^{3}He. The growth equation of the ^{4}He/^{3}He isotope ratio is therefore:

$$\left(\frac{^4\text{He}}{^3\text{He}}\right)_t = \left(\frac{^4\text{He}}{^3\text{He}}\right)_0 + 8\left(\frac{^{238}\text{U}}{^3\text{He}}\right)_t \left(e^{\lambda_{238U} t} - 1\right)$$

$$+ 7\left(\frac{^{235}\text{U}}{^3\text{He}}\right)_t \left(e^{\lambda_{235U} t} - 1\right) + 6\left(\frac{^{232}\text{Th}}{^3\text{He}}\right)_t \left(e^{\lambda_{232Th} t} - 1\right) \quad (3.26)$$

By force of habit, many geochemists continue to use different parameters for isotope system evolution from those just described. For helium, the reverse ^{3}He/^{4}He ratio is very widely used after standardizing it to its value in the atmosphere (1.4×10^{-6}). For other elements, such as neodymium, laboratories employ different standardizations to calculate isotopic abundances. To avoid confusion, it has become common practice to relate isotopic compositions to that of a reference sample, usually the mean of chondritic meteorites, and to use a relative deviation notation, analogous to that already in use for oxygen isotopes. For neodymium, $\varepsilon_{Nd}(T)$ is defined as:

$$\varepsilon_{Nd}(T) = \left[\frac{(^{143}\text{Nd}/^{144}\text{Nd})_{\text{sample}}(T)}{(^{143}\text{Nd}/^{144}\text{Nd})_{\text{chondr}}(T)} - 1\right] \times 10\,000 \quad (3.27)$$

which is the deviation in parts per 10 000 of the ^{143}Nd/^{144}Nd ratio in the sample relative to that of chondrites of the same age T. In a similar way, $\varepsilon_{Hf}(T)$ can be defined for the ^{176}Hf/^{177}Hf ratio.

Chapter 4
Element transport

The theory of element transport is a way of representing the spatial changes in geochemical properties in various contexts, such as movement in the ocean or mantle, the migration of geological fluids or magmatic liquids within a rock matrix, or the attainment of chemical and isotopic equilibrium among minerals within the same rock, etc. It is, in fact, a set of rather complex theories involving some heavyweight mathematics, which we can only touch lightly upon in this book.

The essential concepts forming the core of this theory are those of conservation, flux, sources, and sinks. A conservative property is additive and can only be altered by addition or subtraction at the system boundaries or by the presence of sources and sinks. Mass or number of moles are conservative properties; concentration is not: if a mole of salt is added to a solution already containing one mole, the resulting solution will contain two moles, however the salt is added and whatever the volume of the system. In contrast, two solutions of one mole per liter combine to form a solution of one mole per liter. A flux is a quantity of something (mass, moles, energy, etc.) crossing a unit surface per unit time. The most familiar of these fluxes is volume flow, which is quite simply the velocity v (in cubic meters per square meter per second). Mass flux is the mass content of volume flow, i.e. ρv, where ρ is the density of the medium. The flux of an element i is $\rho v C^i$, where C^i is its local concentration in kilograms (or moles) per kilogram. If the flux of a compound or an element changes suddenly at one point, then a source or sink of this compound or element is present: generally, a chemical reaction or radioactive process is responsible for this.

Transport from one point to another is either by advection or by diffusion. Think of fish in a river. The general motion of the water ensures advective transportation of the fish; the movement of the fish relative to the water is diffusive transportation. Generally, advection is transportation of elements at the speed of the ambient medium, and diffusion of an element is its transportation relative to the ambient medium.

The principles from which the transportation equations derive are difficult to formulate in mathematical terms, but they are easy enough to understand:

- Conservation of total matter, which produces the continuity equation.
- Conservation of element i

$$\frac{\text{variation of mass of } i}{\text{per unit of time}} = \sum \text{flux of } i \text{ across the surface}$$
$$+ \sum \text{sources and sinks} \tag{4.1}$$

In just the same way as we can count our fish from a bridge, from a boat drifting with the current, or from a motor boat, the terms of this equation can be expressed relative to a fixed, or moving, arbitrary reference point.

4.1 Advection

Advection is bulk transport, which is easier to understand in one dimension. Let us consider a medium of density ρ moving at speed v. If, at a point x and over an area A, we consider a slice of matter of thickness l, the balance of variation of mass of i per unit time in this slice is equal to the difference between the incoming and outgoing fluxes:

$$A \, l\rho \frac{\mathrm{d}C^i(x)}{\mathrm{d}t} = A\rho v C^i(x) - A\rho v C^i(x+l) \tag{4.2}$$

or, if l tends toward 0:

$$\frac{\partial C^i}{\partial t} = -v \frac{\partial C^i}{\partial x} \tag{4.3}$$

where the partial derivatives are used to show relative to which variable derivation is applied. If we imagine an observer who, like the boat drifting on the river, moves with the speed v of the medium, the apparent speed of the observer relative to the medium cancels out and the right-hand side disappears. Again, we find the condition that local concentration is invariant: this is known as the Lagrangian representation of the variation. This amazingly simple principle is the one applied in the study of convection of the mantle or the ocean, and in the dispersion of what are known as passive tracers, i.e. tracers not involved in reactions (Fig. 4.1): if we know the field of velocities, we can determine the motion of any point of the medium between two instants and therefore track the position of an atom of element i over time.

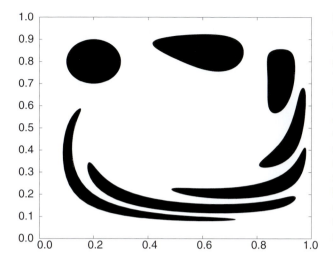

Figure 4.1: Advective dispersion of passive tracers in a velocity field of the x (abscissa) y (ordinate) plane. This diagram shows the evolution of an initial mass of tracers injected at the top left of the system at time t and which follows the clockwise motion of the material (advection). This may represent the evolution of a block of oceanic crust caught up in mantle convection or the dispersion of pollutants released into the ocean.

4.2 Diffusion

Diffusive transfer is transfer over short distances caused by the thermal agitation of atoms or the turbulence of the medium. Let us consider the one-dimensional migration of an element i along the x-axis. Let P be the probability of an atom jumping per unit time from its original site and suppose that it can only jump from one site to an adjacent site at a distance l. A site loses atoms with the same probability to the site to its left as to the site to its right, but only gains half of the atoms lost from each adjacent site. The balance can be written as:

$$\frac{\partial C^i(x)}{\partial t} = \frac{P}{2}C^i(x-l) + \frac{P}{2}C^i(x+l) - PC^i(x) \qquad (4.4)$$

By expanding $C^i(x+l)$ and $C^i(x-l)$ to the second order, we obtain:

$$C^i(x+l) = C^i(x) + l\frac{\partial C^i(x)}{\partial x} + \frac{l^2}{2}\frac{\partial^2 C^i(x)}{\partial x^2} + \mathcal{O}(l^3) \qquad (4.5)$$

$$C^i(x-l) = C^i(x) - l\frac{\partial C^i(x)}{\partial x} + \frac{l^2}{2}\frac{\partial^2 C^i(x)}{\partial x^2} - \mathcal{O}(l^3) \qquad (4.6)$$

Adding these two equations, ignoring terms of order higher than two, and transferring the result into (4.4), gives:

$$\frac{\partial C^i}{\partial t} = \frac{Pl^2}{2}\frac{\partial^2 C^i}{\partial x^2} = D^i\frac{\partial^2 C^i}{\partial x^2} \qquad (4.7)$$

where the diffusion coefficient $D^i = Pl^2/2$ is expressed in square meters per second. The probability P is proportional to: (a) the number of attempts per unit time, i.e. the frequency of vibrations ν of the atom around its resting position; and (b) the proportion of atoms with enough

Figure 4.2: Arrhenius plot of the diffusion coefficients of various elements in different media. Semi-logarithmic graph. Notice temperatures are in K at the bottom and °C at the top.

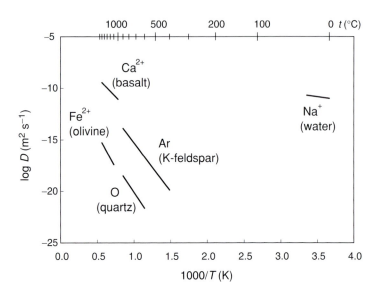

energy to cross an energy barrier E^i, known as the activation energy, dependent on the nature of the atom and the atomic configuration of the site. From Boltzmann's law, this latter fraction is equal to $\exp(-E^i/RT)$ and the temperature dependence of the diffusion coefficient is therefore expressed:

$$D^i = D^i_0 \exp\left(-\frac{E^i}{RT}\right) \tag{4.8}$$

D^i_0 contains the vibrational term. If we introduce the logarithm of D^i as a function of the inverse $1/T$ of absolute temperature (Arrhenius plot), we obtain a straight line of slope $-E^i/R$. A few values of the diffusion coefficient in water, magmatic liquids, and minerals at magmatic temperatures are shown in the Arrhenius plot in Fig. 4.2.

The physical meaning of the diffusion coefficient can also be understood in a different way. Equation (4.7) can be re-written as follows (quite analogous to (4.1)):

$$\frac{\partial C^i}{\partial t} = -\frac{\partial J^i}{\partial x} \tag{4.9}$$

where the diffusive flux J^i of species i is defined as:

$$J^i = -D^i \frac{\partial C^i}{\partial x} \tag{4.10}$$

This shows that the transport of i by diffusion is proportional to the concentration gradient $\partial C^i/\partial x$ and that the diffusion coefficient D^i is the constant of proportionality. The minus sign ensures that the flux is directed toward the part of the system with a lower i concentration. Equation (4.10) is Fick's law.

For example, we can calculate the diagenetic diffusive flux of manganese ($i = Mn$) in sedimentary pore water to ocean bottom water. Let us assume that, because of its reducing character, interstitial water 10 cm beneath the surface has a manganese concentration of 10 µmol l^{-1}. Seawater, being highly oxidized, has a manganese concentration that is virtually nil. The diffusion coefficient of manganese in seawater is $D^{Mn} = 0.7 \times 10^{-9}$ m^2 s^{-1}. The local diffusive flux of manganese to the ocean in absolute terms is therefore:

$$J^{Mn} = 0.7 \times 10^{-9} \frac{10 \times 10^{-6} \times 10^{-3} - 0}{0.1} = 0.7 \times 10^{-16} \text{ mol m}^{-2} \text{ s}^{-1}$$

Clearly, as we considered the probability of a forward or backward jump as equal (Brownian motion), the mean displacement of an atom is zero. Nonetheless, we can calculate a measure of the dispersion of a population of atoms initially in position x through another simple statistical concept, the standard deviation of displacement, i.e. the mean quadratic displacement. During time t, the number of jumps by an atom is equal to Pt and the quadratic distance travelled by an atom is Ptl^2, or $2D^i t$. The mean quadratic deviation, i.e. the distance δ over which two-thirds of the atoms will be found on either side of their initial position, is therefore $\delta = \sqrt{2D^i t}$. This distance is a measure of the migration characteristic of the atoms under the effect of diffusion and is generally short; it is typically a few centimeters per year for an ion in seawater, meters per million years for the Fe–Mg exchange of olivine in the mantle, or a micrometer per million years for oxygen isotope exchange in metamorphic quartz.

Diffusion is not therefore a process for long-distance transport, but it usually does allow minerals to achieve equilibrium with each other or with solutions percolating through the rock. The time required for a spherical mineral of radius a to lose a fraction F of an element it initially contained complies fairly well with the equation:

$$F = 1 - 6\sqrt{\tau/\pi} + 3\tau \qquad (4.11)$$

where $\tau = D^i t/a^2$ is a scalar proportional to time known as dimensionless time. This equation is widely utilized for calculating diffusion coefficients; for example, for rare gases, by evaluating the rate of degassing of a particular mineral at a certain temperature as a function of time, or for soluble compounds, by immersing them in a liquid.

When advection and diffusion are both operative at the same time, their relative importance can be evaluated with the Peclet number Pe, with Pe $= vd/D^i$, where d is a characteristic dimension of the system

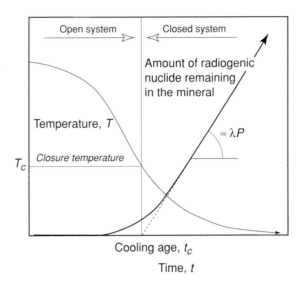

Figure 4.3: Cooling age and closure temperature of a chronometric system with decay constant λ. The cooling time and the corresponding cooling age t_c are defined by linearly extrapolating the ingrowth of radiogenic nuclides at zero, while the closure temperature T_c is the temperature of the mineral at this time. The system is open to the loss of the daughter isotope for $T > T_c$ and closed for $T < T_c$. The slope of the daughter-isotope evolution curve is simply the activity λP.

(e.g. grain size or conduit diameter). Diffusion predominates when Pe < 1, and advection for Pe > 1.

4.2.1 Closure temperature of radioactive chronometers

Diffusion theory can be applied to calculate the resistance of radioactive chronometers to thermal disturbances or for evaluating the equilibration temperature of a mineral assemblage (thermobarometry). A number of important applications in geochronology rely on the so-called concept of closure temperature defined for radiogenic isotopes by Dodson. The high activation energies of diffusion in solids (Eq. (4.8)) cause diffusion coefficients to vary closely with temperature. The transition from open system to closed system therefore occurs in a narrow range of temperature (Fig. 4.3). Above a certain critical temperature T_c and therefore for a diffusion coefficient greater than D_c^i, where:

$$D_c^i = D_0^i \exp\left(-\frac{E^i}{RT_c}\right) \qquad (4.12)$$

the system can be considered open. The radiogenic isotope will only begin to accumulate below T_c, the temperature at which the stop-watch is started, with the transition occurring within a few tens of degrees. Let us see how to evaluate T_c for a given isotope in a given mineral. From an equation such as (4.11), we see that the critical parameter of the loss process is dimensionless time $\tau = D^i t/a^2$. Let us decide that the system is open for $\tau > \tau_c$ and closed for $\tau < \tau_c$, where τ_c is a constant factor dependent on geometry (0.02 for a sphere and 0.12 for a sheet). If the significance of the radius a of the mineral is clearly perceived, what is the time characteristic to be introduced in τ? Dodson suggests that a characteristic time is the time $t = \theta$ required for the diffusion coefficient to reduce by a factor e, and is equal to:

$$\frac{1}{\theta} = -\frac{d \ln D^i}{dt} = \frac{E^i}{R} \frac{d(1/T)}{dt} = -\frac{E^i}{RT^2} \frac{dT}{dt} \qquad (4.13)$$

This time can be estimated from experimental data on diffusion coefficients and even approximate knowledge of the local rate dT/dt of cooling after formation of the minerals under analysis. Let us define the dimensionless time τ_0 as:

$$\tau_0 = \frac{D_0^i \theta}{a^2} \qquad (4.14)$$

Dividing $\tau_c = D_c^i \theta / a^2$ by τ_0, we obtain:

$$T_c = \frac{E^i}{R \ln (\tau_0/\tau_c)} \qquad (4.15)$$

By introducing the value θ given by (4.13) and (4.14) for $T = T_c$ into τ_0, we can now evaluate T_c by trial and error and iteration. For example, let us consider the closure of potassium feldspar crystals, assumed for simplicity to be spherical with a radius of 2 mm, to the diffusion of ^{40}Ar produced by decay of ^{40}K. We assume that argon diffusion is described by the following parameters: $D_0 = 1.4 \times 10^{-6}$ m^2 s^{-1}, $E^i = 180\,000$ J mol^{-1}, $a = 0.002$ m, $dT/dt = 5$ K My^{-1}. We would normally have to find the closure temperature by trial and error and we would converge to $T_c = 615$ K. Upon inserting $T = 342\,°C = 615$ K into (4.13), we obtain $\theta = 1.10 \times 10^{14}$ K, which corresponds to $\tau_0 = 3.85 \times 10^{13}$ (Eq. (4.13)). Inserting these values into (4.15), we can check that $T = T_c$. We conclude that crystals of potassium feldspar with a radius smaller than 2 mm start losing their radiogenic argon when heated above 342 °C, or, equivalently, that they start being retentive for argon as soon as ambient temperature drops below 342 °C. Closure temperature depends mostly on grain size. Error bars on closure temperatures are in the order of 50–80 °C.

Closure temperatures of different chronometers for specific minerals can be calculated from diffusion experiments. Figure 4.4 shows a hypothetical curve constructed to show cooling after the formation of mountain ranges in a given region. If equilibration pressure is known, all of this information forms the pressure–temperature–time (PTt) path of exhumation of the orogen, an essential tool in modern tectonics. This same tool is also much used in petroleum exploration to determine the thermal evolution of sedimentary basins and the probability of temperatures suitable for oil formation being attained locally.

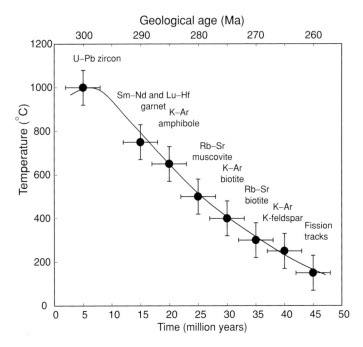

4.2.2 Other applications

The diffusion equation, with or without an advection term, is involved in
the description of many other geochemical processes. A very important
case is that of early diagenesis, a set of reactions occurring a few meters
beneath the sea floor whereby mud is transformed into sediment. An
observer in a fixed position at the water–sediment interface "watches"
the sediment sink at the rate of sedimentation and must describe the
chemical processes in an advective context. The diffusion of elements in
interstitial water and the reactions between that water and the sedimented
particles control the nature of mineralogical transformations and the flux
back into seawater or diagenetic liquids as the sediments become compacted. Of particular importance are the reduction reactions of nitrates
to nitrogen and of sulfates to sulfides by which many micro-organisms
oxidize sedimented organic matter. These processes lie behind the dissolution of manganese nodules and the formation of sedimentary pyrite
(fool's gold), in particular.

Diffusion is also involved in the growth of minerals from solutions
or magmas: an olivine crystal that grows from a basalt is surrounded by
a boundary layer depleted in the essential components of this mineral,
notably Mg. This layer can often be observed under the microscope.
The diffusion of these components allows re-supply and, consequently,
the continued growth of the mineral. In the mantle, homogenization of

compositions at distances greater than about ten meters is thought to be achieved by mechanical stirring caused by convection. Conversely, diffusion ensures homogenization over short distances, in the order of a few centimeters.

4.3 Chromatography

Imagine, on a hot summer day, a parade stretching along a long avenue lined with cafes and shady terraces. It is likely that some of the participants from time to time will succumb to the temptation of enjoying a few minutes of refreshment before joining the parade again, and that some will even take a liking to these halts. We can guess what our parade will look like after a few hours: the head of the procession will be reduced to the virtuous and sober participants; while behind, the atmosphere will be enlivened by all those who have made frequent stop-offs at the terraces. This type of mechanism is the essence of chromatographic separation.

In analytical chemistry, chromatography is a technique of element separation that involves the percolation of a liquid (the eluent or mobile phase) through a porous matrix (the stationary phase composed, for example, of ion-exchange resin) and exchange elements with this matrix. The variable affinity of the stationary phase for the elements lets each of them percolate through the matrix with a different velocity, which eventually ensures their separation. In rocks, similar processes take place when geological fluids move through the pores of a sediment during diagenesis or through rock layers during metamorphism: because of the large volumes of fluid involved and because of the broad range of reactivity from one element to another, considerable chemical separation in the fluid and strong mineralogical and geochemical modifications of the rock matrix are commonplace.

In a two-phase flow, such as water percolating along an aquifer through a porous soil or a magmatic liquid in a molten rock matrix, or of particles sedimenting in the ocean, the processes are both more complicated and more diverse. The constitutive equations of chromatography are difficult to establish because, besides continuity and conservation equations for each species in each phase (e.g. solid and liquid), we have to add a condition describing transfers between phases, such as phase change kinetics (melting, dissolution, or crystallization) and chemical fractionation. The concentration of an element i in the interstitial mobile phase (liquid) can generally be described by a balance statement of the type:

$$\frac{dC^i_{liq}}{dt} = \text{diffusion} + \text{advection} + \text{phase changes} \qquad (4.16)$$

To illustrate the concepts in the simplest way, let us ignore the diffusive term and the term related to transfer during phase change. The advective effect is then dominant. Let us denote v_{liquid} the velocity at which the liquid moves relative to the solid matrix. The theory of chromatography is quite heavy going and we will have to accept the simple result as a close approximation; the velocity v^i at which an element i moves is given by:

$$v^i \approx \frac{\varphi}{\varphi + D^i(1-\varphi)} \, v_{\text{liquid}} \tag{4.17}$$

where φ is porosity and D^i the partition coefficient of the element i between the solid matrix and the liquid. Ions that migrate with the water of an aquifer without interacting with the impregnated rock ($D^i \approx 0$) will therefore move faster than the ions that readily exchange between rock and water ($D^i \gg 0$) and spend a large part of their time of transit in the solid rock. For instance, chlorine ions have no affinity for minerals and, like a dying substance, move at the same speed as the water. Ions such as Zn^{2+} or Pb^{2+} are readily absorbed by the surface of the minerals. They are very much delayed by this reactivity. This is the principle behind the purification of natural water in the ground. The containment within nuclear waste repositories of radioactive actinides, which normally have $D^i_{\text{mineral/water}} \gg 0$, also relies heavily on this property.

Let us now consider the situation where the partition coefficient D^i of an element i and, therefore, its speed in the interstitial mobile phase, depends on its concentration. Suppose that D^i decreases when the concentration in i increases. In this case, the greater the concentration of the element, the less reactive it will be. By applying (4.17), it can be seen that the ions of highly concentrated zones will catch up with those of less concentrated zones (Fig. 4.5), thus forming what are known as metasomatic fronts or even more complicated geometric patterns. The description of magma migration is far more complex still because,

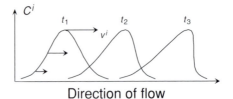

Direction of flow

Figure 4.5: Principle of chromatographic fractionation. Different ions propagate along the column at different speeds. Each curve represents the concentration profile of a same element at different times ($t_1 < t_2 < t_3$). If the speed of propagation of an element i in the porous medium depends on its concentration C^i in the liquid and partition coefficient D^i, concentration fronts rise and collapse.

contrary to the ground water table where the water itself reacts very little with the rock matrix, all the major elements of the magma exchange intensely with the solid residue.

Chromatography theory applies to other cases of two-phase transport. An important concept in oceanography is the entrainement – or scavenging – of elements by particles such as dead organisms, waste matter, or atmospheric dust. In this case, the particles take the place of the liquid occupying porosity φ of the general theory, while seawater plays the role of the rock matrix. What conditions reactivity is the seawater–particle partition coefficient D^i of elements between seawater and particulate matter (note the direction of exchange, which gives reactivity a counter-intuitive character). Elements that are not reactive with particles ($D^i \approx \infty$), such as sodium and chlorine, remain motionless and are not affected by particle flow. Elements subjected to intense entrainment by solids ($D^i \approx 0$), such as lead, cadmium, and rare-earths, move down the water column almost as quickly as the particles and are soon evacuated from the system.

4.4 Reaction rates

Let us consider the reaction by which marine sulfates are reduced at the water–sediment interface by accumulated organic matter, which we will represent by the very simple chemical formula CH_2O (the precise formula is immaterial to the explanation). This reaction produces hydrogen sulfide, which will precipitate as pyrite FeS_2 with the ambient iron, and a bicarbonate ion from the reaction:

$$2CH_2O + SO_4^{2-} \rightarrow H_2S + 2HCO_3^- \qquad (4.18)$$

By indicating the molar concentrations of the species with square brackets, the conservation of matter during the reaction imposes the following conditions of evolution (see Appendix C):

$$-\frac{d\,[CH_2O]}{2} = -\frac{d\left[SO_4^{2-}\right]}{1} = +\frac{d\,[H_2S]}{1} = +\frac{d\left[HCO_3^-\right]}{2} = d\xi \quad (4.19)$$

where ξ is the degree of advancement of the reaction. The quantity $d\xi/dt$ is the reaction rate. This depends on the probability that the reactants will collide and therefore, on their concentration, and, as with a diffusion coefficient, on the frequency ν of vibrations and on the potential barrier ΔE to be crossed so as to "activate" the reactants. This is therefore written:

$$\frac{d\xi}{dt} = k\left[CH_2O\right]^2\left[SO_4^{2-}\right] \qquad (4.20)$$

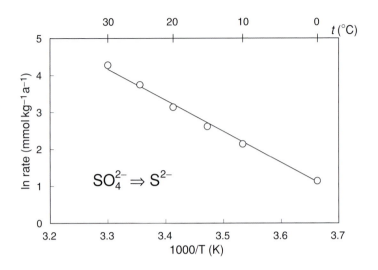

Figure 4.6: An example of reaction kinetics: sulfates being reduced to sulfides in diagenetic solutions. A linear relation is found in the semi-logarithmic Arrhenius plot (data from Westrich and Berner, 1984).

where it can be seen that the probabilities are multiplicative properties, with

$$k = k_0 \nu \exp\left(-\frac{\Delta E}{RT}\right) \qquad (4.21)$$

where k_0 is a constant that includes many parameters, notably the specific surface area of the reactants (a finely divided mineral is more reactive than a single large crystal). We recognize a Boltzmann law, which can be represented in an Arrhenius plot $\ln k = A + B/T$ (Fig. 4.6). A large number of kinetic constants were determined in the 1980s and 1990s. At high temperatures, reaction kinetics is normally fast but, in the context of diagenesis and weathering, dissolution is slow.

Reference

Westrich, J. T. and Berner, R. A. (1984). The role of sedimentary organic matter in bacterial sulfate reduction. *Limnol. Oceanog.*, **29**, 236–249.

Chapter 5
Geochemical systems

This chapter looks at the changes that over time affect the geochemical properties of a system or a set of systems, such as the mantle, the crust, or the ocean, when subjected to disturbances whether caused naturally or by human activity. The essential concepts utilized – residence time and forcing – are taken from chemical engineering. Viewing system Earth as a chemical factory composed of reactors, valves, sources, and sinks, has proved to be a simple and robust model. The theory goes by various names, with the "box model" probably the most widely used. We will first set out the principles by describing the behavior of a system with a single reservoir and then go on to generalize the approach.

5.1 Single reservoir dynamics

Let us begin by considering a lake (Fig. 5.1) containing a mass of water M that we will take to be constant. A river flows through the lake with a rate of flow Q, which we will express in kilograms per year. Q is therefore the same upstream and downstream. We are interested in the balance of a chemical species in the lake. A chemical element introduced upstream may be carried away downstream or be taken up by sedimentation, at a rate P also expressed in kilograms per year. The lake itself is considered homogeneous, being well mixed by turbulent flow and by convection.

The concentration C^i of an element i is therefore the same in the lake and in the downstream outflux. This balance can be written:

$$M\frac{\mathrm{d}C^i}{\mathrm{d}t} = QC_{\mathrm{in}}^i - QC^i - PC_{\mathrm{sediment}}^i \tag{5.1}$$

Figure 5.1: Simple box model (a single reservoir) typical of a lake. M: mass of water in the lake, Q: water inflow = outflow, P: sedimentation rate.

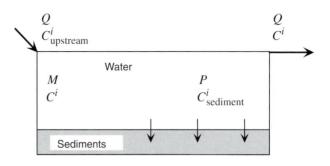

and, as usual, we will introduce a partition coefficient D^i to allow for fractionation of the element between solid and water, such that $C^i_{\text{sediment}} = D^i C^i$. Rearranging this gives:

$$\frac{\tau_{\text{H}}}{1 + \beta^i} \frac{dC^i}{dt} + C^i = \frac{C^i_{\text{in}}}{1 + \beta^i} \qquad (5.2)$$

where $\tau_{\text{H}} = M/Q$ is the time to renew the water in the lake and $\beta^i = PD^i/Q$ is the reactivity of the element relative to sedimentation. This reactivity is zero for an inert tracer, such as a coloring agent or chlorine ion, and increases with the solid/liquid partition coefficient of the element. The term in front of the derivative is the residence time $\tau^i = \tau_{\text{H}}/(1 + \beta^i)$ of element i in the lake, i.e. the mean time that atoms of i spend in the lake before being removed (see Appendix H).

The form of (5.2) is extremely instructive. Time is now an integral part of the balance, contrary to the distillation equations mentioned earlier. If the concentration of i upstream is zero, we obtain by integration the change in a lake initially at concentration $C^i(0)$, possibly by accidental spillage of a pollutant at $t = 0$:

$$C^i(t) = C^i(0)e^{-t/\tau^i} \qquad (5.3)$$

The lake is said to relax from its initial state described by the concentration of the element at $t = 0$. The measurement of relaxation rate is the residence time τ^i, which is shorter for more reactive elements and less than the time it takes to renew the lake water. The residence time τ^i indicates how quickly a disturbance in the balance of element i is damped out by the lake.

When $t \gg \tau$, relaxation is complete, the derivative in (5.2) vanishes and a constant concentration is reached. This state is known as steady state. The simple relation $C^i(\infty) = C^i_{\text{in}}/(1 + \beta^i)$ shows that the right-hand side is the concentration when the initial condition is fully relaxed (Fig. 5.2). This is called the forcing term. The forcing term, here the input from upstream, shows in which direction the system is heading.

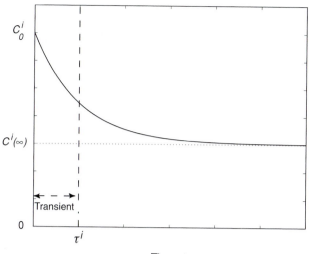

Figure 5.2: Relaxation of a perturbation at time $t = 0$ of the concentration in the element i within a single box system. The residence time τ^i measures the rate at which the perturbation is relaxed. The transient is the regime in which the initial perturbation can still be perceived.

Again it can be seen that the final concentration of the lake is lower when the element is more reactive.

The reader can check from (5.2) that the residence time of element i is simply the ratio of the mass contained in the reservoir to the output of this element at steady state, i.e.:

$$\tau^i = \frac{MC^i(\infty)}{QC^i(\infty) + PC^i_{\text{sediment}}(\infty)} = \frac{\text{mass of element } i \text{ in the reservoir}}{\text{output rate of element } i} \quad (5.4)$$

Since the system is at steady state, the output rate may conveniently be replaced by the input rate. The residence time of elements in the ocean may thus be evaluated from data of concentration in seawater and rivers in Appendix A and geophysical data in Appendix G. For example, the residence time of strontium is $(1.4 \times 10^{21}$ kg $\times$ 8000 µg l^{-1}) / $(3.6 \times 10^{16}$ kg a$^{-1} \times$ 70 µg l^{-1}), or 4.4 million years.

More complete and important expressions are given in Appendix H.

A fundamental aspect of reservoir dynamics is that the more reactive pollutants will be eliminated faster and their residual concentration will be lower. Reactivity is a function of two factors: the rate of sedimentation (sediment cleans the water) and the fractionation of the element between solid and liquid. Hence the apparent paradox: elements that poison micro-organisms and sediment are eliminated faster.

This elementary theory explains some common but essential aspects of the geochemistry of large geological systems. Introducing the definition of τ^i and $C^i(\infty)$ into (5.2) and rearranging, we get:

$$\frac{1}{C^i - C^i(\infty)} \frac{d\left[C^i - C^i(\infty)\right]}{dt} = \frac{d\ln\left[C^i - C^i(\infty)\right]}{dt} = -\frac{1}{\tau^i} \quad (5.5)$$

which indicates that the relative deviations of element concentrations from their steady-state values fluctuate at a rate scaled by the inverse of the element residence time. Inert elements, i.e. elements whose partition coefficients D^i are very small, have long residence times. Their concentrations therefore remain very stable. These elements are highly concentrated (i.e. they are major elements), their residence time is long, and they display no fast temporal variations. This is the case of strontium in the ocean, for which we have calculated a residence time of 4.4 Ma. By contrast, reactive elements with high partition coefficients D^i and therefore short residence times are very quickly depleted in the reservoir (i.e. they are minor elements). Their short residence time means they readjust quickly after a disturbance, but their concentration level fluctuates wildly about steady state. To use an analogy, a reservoir such as a lake, the ocean, or the mantle, behaves like a low pass filter, eliminating fluctuations in concentration with shorter time characteristics than the residence time of the element. If a time characteristic τ_m of mixing in the reservoir can be defined and the residence time τ^i of an element i is such that $\tau^i > \tau_m$, this element will reside long enough in the reservoir to be well homogenized. If its residence time is shorter than the reservoir mixing time, the element will be drained away in the outflow before it is effectively mixed and will be distributed heterogeneously around the system. For instance, concentrations of Sr (8 ppm) and ^{87}Sr/^{86}Sr (0.7093) are homogeneous across the entire ocean.

In the ocean, the export process of most elements is sedimentation and the partition coefficient measures sediment particle/seawater fractionation. In the mantle, the export process is magmatism and the coefficient D^i that measures the magma/residual solid fractionation is the inverse of the partition coefficient usually considered by petrologists. Inert elements or ions are residual, very homogeneous, abundant, and of invariable concentration: this is the case of Mg, Ni, and Cr in the mantle, of Cl, Na, and SO_4 in the ocean, and of N_2, O_2, and Ar in the atmosphere. In contrast, reactive elements are normally at the trace level and exhibit wide variability in space and time: this is the case of the incompatible lithophile elements (Th, Ba, La, Rb, etc.) in the mantle, of Pb, Cd, Zn, and La in the ocean, and of methane and ozone in the atmosphere.

In the absence of any input from upstream, (5.2) can be reformulated as:

$$-\frac{dC^i}{C^i dt} = \frac{1}{\tau_H} + \frac{PD^i}{M} = \frac{1}{\tau_H} + \frac{1}{\tau_S^i} \tag{5.6}$$

This equation expresses that the total probability of an atom i leaving the system per unit time is the sum of two probabilities: the probability of i exiting downstream, which is equal to the inverse of the renewal time of the water, plus the probability of i being lost to the sediment, which is the inverse of removal time by the solid $\tau_S^i = M/PD^i$. As a simple but powerful rule, the probabilities of removal through various routes are independent and therefore additive and so are the inverse residence times. We will see below that a third removal mechanism is radioactivity and that the probability of removal by this process is simply the decay constant λ.

A third aspect of residence time is that of the distribution of the "ages" of atoms within the reservoir. To understand this idea, let us take another analogy, namely the reservoir of human populations. Suppose, for simplicity, that the number of births every year is constant. A healthy population is a poorly "mixed" population since the exit, i.e. death, draws off the elderly preferentially. A trip around a graveyard might show that the average age of death is about 70–80 years, the mean span of our time here below. However, the average age of the population is much younger, probably about 30–40 years. Let us imagine now, an outbreak of a virus that affects old and young indiscriminately. The population becomes well mixed in the sense that the probability of exiting the system is independent of the individual's age. The gravestones would then show an average age of demise close to the mean age of the population. This latter case is the one most often considered in the dynamics of natural systems where the probability that a particle leaves the system does not increase with the particle's age. It can be shown (Appendix H) that particles in a well-mixed system have an exponential age-group histogram (exit is a Poisson process). In a well-mixed reservoir, not all the individuals have the same residence time but they all have the same probability of exit, regardless of their previous history in the system. Such an example of age distribution is given by the survival of clastic sediments since the end of the Paleozoic (Fig. 5.3). The proportion of sediments surviving beyond a given age decreases exponentially with time. The average age of the sediment is equal to its mean survival time of about 150 million years. This demonstrates the stationary character of the total mass of clays, sandstones, etc.

The concept of residence time requires that fluctuations in the input of an element of duration shorter than its residence time do not show in the output. It has very diverse applications in oceanography and limnology as well as in magmatic petrology or mantle geochemistry. Let us take an example from the magmatic field. An erupting magma reservoir that is continuously replenished can be thought of in the same way as a lake, the upstream being the inflow of fresh magma, the downstream the eruptive activity, and the cumulates the equivalent of sediments. In this context, the compatible elements that are entrained by crystallization are reactive elements and their residence time in the reservoir is shorter than that of the incompatible and therefore inert elements. A reservoir like

Figure 5.3: Age distribution
of fine clastic sediments (after
Mackenzie and Garrels, 1971).
The straight line arrangement
implies that the sedimentary
reservoir is well mixed with a
probability of erosion
independent of sediment age.
The slope of the line indicates
that this type of sediment
survives on average about
150 million years before being
eroded.

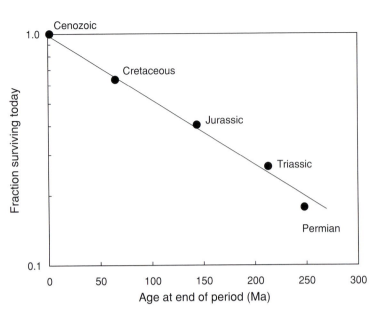

this will therefore damp out all fluctuations in the concentration of an element whose time characteristic is less than its residence time. The characteristic period of fluctuation of the Ce/Yb ratio in the lava of the active volcano Piton de la Fournaise (on the island of Réunion) is of a few years (Fig. 5.4). As both these elements are incompatible ($\beta^i \approx 0$), we thus have an important constraint on the renewal time τ_H of magma in the reservoir supplying the volcano. This time τ_H is of the order of a few years to a few decades. Given the eruption rate Q, the mass $M = Q \times \tau_H$ of the reservoir can be evaluated. These simple observations limit the volume of the magma reservoir to less than one cubic kilometer. The concept of residence time has greatly altered the way we see volcanic systems: until a few years ago, most volcanoes were thought to tap immense magma chambers, which turned out to be much smaller systems of feeder dikes filled with crystal mush.

If the element in question is radioactive with a decay constant λ_i, the radioactive decay $\lambda_i M C^i$, can be subtracted from (5.1), modifying (4.2) as:

$$\frac{\tau_H}{1 + \beta^i + \lambda_i \tau_H} \frac{dC^i}{dt} + C^i = \frac{C^i_{in}}{1 + \beta^i + \lambda_i \tau_H} \tag{5.7}$$

This equation applies, in particular, when determining the mixing time of the ocean. Because the deep ocean is isolated from the atmosphere, the radioactive carbon-14 it acquired during its time at the surface is no longer renewed and so decays within a closed system. Oceanographic measurements indicate that the $^{14}C/^{12}C$ ratio of deep water makes up

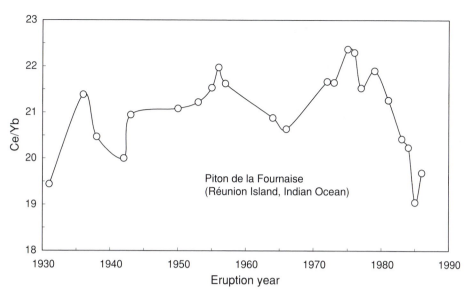

Figure 5.4: Ce/Yb ratio in recent lavas of the Piton de la Fournaise volcano (island of Réunion). The rapid fluctuation of this ratio in a few years indicates that the magmatic reservoir where the fractionation of the two elements occurs contains a volume of lava equivalent to just a few years' eruption. There is therefore no magmatic chamber larger than a cubic kilometer beneath this volcano (after Albarède and Tamagnan, 1988).

only 84% of the same ratio in the surface water. Suppose that the ocean composition has reached steady state $(dC^i/dt = 0)$. Let us arbitrarily divide the ocean into two reservoirs, one of deep water and one of surface water, which exchange water by vertical mixing. The inflow into the deep reservoir is therefore the downwelling of surface water. Dividing (5.7) for ^{14}C by (5.2) for the stable ^{12}C, we get:

$$\left(\frac{^{14}C}{^{12}C}\right)_{deep} = \frac{1 + \beta^C}{1 + \beta^C + \lambda_{14C}\tau_H} \left(\frac{^{14}C}{^{12}C}\right)_{surface} \tag{5.8}$$

The term for chemical reactivity β^C, which is identical for both carbon isotopes, is small enough compared with unity to be ignored. The reader will check that by introducing the relative values of isotope ratios given above and the value of λ_{14C} from Table 3.1, the renewal time for deep seawater τ_H can be calculated at 1600 years.

5.2 Interaction of multiple reservoirs and geochemical cycles

The dynamics of a multiple reservoir system differ from those of a single reservoir in different respects. Collective readjustment is invariably faster than it would be if the reservoirs were considered separately. We will see that ignoring this principle would lead to very serious errors.

Let us consider the case of a change in the number of atoms of an element in two reservoirs that exchange matter with one another, for example the set of the "surface water" and "deep water" oceanic

reservoirs exchanging material by vertical advection, or of the mantle–
continental crust system. Let us call $p^{1\to2}$ the probability for an atom
to be transferred from reservoir 1 to reservoir 2 in the unit time, with a
similar definition for $p^{2\to1}$. If n_1 and n_2 stand for the number of moles
of the element in question in each reservoir, we get the two equations:

$$\frac{dn_1}{dt} = -p^{1\to2}n_1 + p^{2\to1}n_2 \tag{5.9}$$

$$\frac{dn_2}{dt} = p^{1\to2}n_1 - p^{2\to1}n_2 \tag{5.10}$$

Let M_1 and M_2 be the masses of the two reservoirs and Q the mass of
material (water, magma) exchanged between them per unit time. As in
the previous section we can relate the probability of transfer to the ratio
of the flux of element to its amount in the reservoir, for instance:

$$p^{1\to2} = \frac{1}{\tau_1} = \frac{Q(n_1/Q)}{M_1(n_1/Q)} \tag{5.11}$$

where concentrations are represented by n/M in parentheses and τ_1 is the
residence time of the element in reservoir 1. The balance can therefore
be written as two conservation equations:

$$\frac{dn_1}{dt} = -\frac{QD^{1\to2}}{M_1}n_1 + \frac{QD^{2\to1}}{M_2}n_2 \tag{5.12}$$

$$\frac{dn_2}{dt} = \frac{QD^{1\to2}}{M_1}n_1 - \frac{QD^{2\to1}}{M_2}n_2 \tag{5.13}$$

In each term we recognize the product of the material flux Q multiplied
by a concentration n_i/M_i and by a partition coefficient $D^{i\to j}$ allowing
for possible fractionation of the element as it passes from reservoir i to
reservoir j. When material is extracted from the mantle as a basaltic
or andesitic magma and incorporated into the continental crust, the
magma is chemically fractionated relative to its starting mantle and
$D^{\text{mantle}\to\text{crust}}$ increases with the lithophile character of the element. By
contrast, when water is exchanged between two parts of the ocean, the
coefficients D may be equal to 1. In terms of residence time τ_1 and τ_2 of
the element in each reservoir, the system becomes simply:

$$\frac{dn_1}{dt} = -\frac{1}{\tau_1}n_1 + \frac{1}{\tau_2}n_2 \tag{5.14}$$

$$\frac{dn_2}{dt} = \frac{1}{\tau_1}n_1 - \frac{1}{\tau_2}n_2 \tag{5.15}$$

It can be seen that these two equations have a zero sum, which reflects
the constancy of the total inventory $n_1 + n_2 = N$ of the element in the

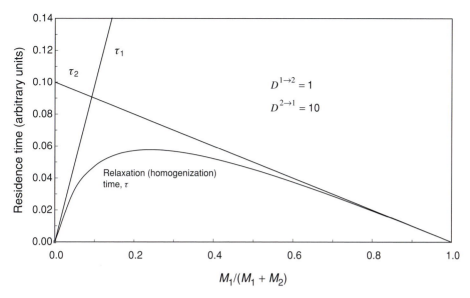

Figure 5.5: Calculation of the homogenization time of an arbitrary element in a two-reservoir system (arbitrary units). Partition coefficients D indicated are those governing the transfer of the element from one reservoir to the other. This homogenization time depends on the relative mass of each reservoir and is less than the time of residence of the element in each reservoir. The cooperative character of exchanges accelerates homogenization compared with what might be expected for readjustment of each reservoir separately.

system. By introducing this condition into (5.14), and rearranging and solving it, we obtain:

$$n_1 = n_{1,0}\, e^{-t/\tau} + N\frac{\tau_1}{\tau_1 + \tau_2}\left(1 - e^{-t/\tau}\right) \qquad (5.16)$$

where $n_{1,0}$ is the initial value of n_1. A symmetrical equation would be obtained for n_2. The overall relaxation time τ of the system is defined by:

$$\frac{1}{\tau} = \frac{1}{\tau_1} + \frac{1}{\tau_2} \qquad (5.17)$$

indicating that τ is shorter than either τ_1 or τ_2 (Fig. 5.5). The steady state is reached and homogenization between the two reservoirs is faster than a simple analysis of the residence time of each of the reservoirs seems to indicate. The falsely intuitive idea that the chemical response time of a system should need longer to adjust than the residence time of its slower reservoir is clearly wrong. The reason for this acceleration can be understood by rewriting (5.14) as:

$$\frac{\tau_1 \tau_2}{\tau_1 + \tau_2}\frac{dn_1}{dt} + n_1 = \frac{\tau_1}{\tau_1 + \tau_2}N \qquad (5.18)$$

and comparing this equation with (5.2). The forcing term arising on the right-hand side from coupling with reservoir 2 interferes with the straightforward relaxation of reservoir 1 and always in the direction of acceleration.

The reasoning just developed applies to more complex sets of reservoirs, but we would then need tools from linear algebra. We have considered only systems without an external forcing effect; such an effect

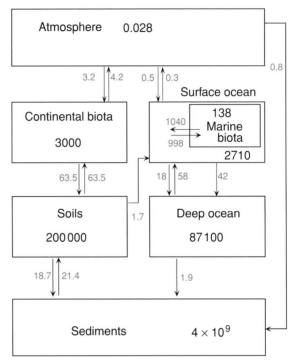

Figure 5.6: The external cycle of phosphorus (Lerman, 1988). Masses (solid type) and fluxes (grey tinted type) are in millions of tons of phosphorus per year.

should, however, be allowed for where river water flows into the ocean or cosmogenic matter into the mantle. In general, this theory underlies the understanding of geochemical cycles, i.e. processes of exchange of each element between the Earth's different reservoirs. The conditions for describing an elementary cycle are: (1) essential reservoirs must be identified and their contents estimated, and (2) fluxes of elements between reservoirs must be evaluated over a sufficient length of time compared with the relaxation time of the system. The phosphorus cycle is given here (Fig. 5.6) as a good example of a cycle of an element taken from the literature.

One aspect of cycles that is difficult to grasp is that of the time scale at which they apply. It is easy to see that estimating the transportation of an element to the ocean by rivers in flood or from the mantle to the atmosphere by extrapolating from observation of a volcanic eruption gives an inaccurate representation of flows. Catastrophic transfer (or pulse transfer) is particularly difficult to fit into the picture. Another condition may seem implicit in the idea of cycle, that of the stationary state, or the balanced cycle: an atom is passed many times through all of the reservoirs and the inventory of each reservoir no longer varies. This condition is always an approximation to some extent. Suppose we are interested in the carbon cycle at appropriate time scales for historical climatic variations

(100–2000 years). Should we include all carbon transfers between the mantle, crust, ocean, atmosphere, and biomass and at all time scales, from diurnal (daily) rhythms of the biomass to fluctuations of volcanic emission, at very long geological scale? At the scale of a century, the effect of diurnal fluctuation is that of the sum of 365×100 cycles and its effect is therefore virtually nil. At the scale of a millennium, the long geological variations will be imperceptible. Adding that effect would be the same as arbitrarily introducing a transfer equation equivalent to a "non information." Now, it is well known in mathematics that adding an equation of the type $0 = 0$ makes an equation system insoluble. To ensure the cycle is described properly, we will therefore choose conservatively to draw up an inventory of processes that are important at time scales of 1 to 100 000 years, ignoring faster processes and treating longer-term ones as forcing to be dealt with by separate parameters. The use of a time "window" adequate for all the processes to be considered is an essential condition for the construction of correct geochemical cycles.

It is legitimate to ask about the relationship between the representation of chemical transfers by the advection–diffusion equation, in which the concentration of an element is represented at each point in space and at each moment in time, and the representation of the multiple reservoir model ("box model") with no spatial indications. Both representations are justifiable and consistent, the two essential differences being that, in the box model, concentration is averaged over the entire reservoir while the fluxes are integrated over the entire area bounding each reservoir. It could thus be shown that by multiplying the number of reservoirs and making their size tend toward zero, a continuous transition can be made from one representation to the other.

5.3 Mixing

Let us now try to describe how chemical heterogeneities are destroyed by the convective movements of the mantle or the ocean. Anyone can imagine the fate of a packet of different colored sticks of play dough left in the hands of a small child from an early stage where the sticks are distorted but remain identifiable to a homogeneous bluish-gray dough, passing through a marbled stage with multicolored stripes stretched and folded back on themselves many times (Fig. 5.7). Obviously if a representative sample of the average of the packet is to be taken, it must be sufficiently large compared with the thickness of the stripes. The concept of homogeneity is therefore dependent on the scale of observation or of sampling, and it will always be possible to spot local heterogeneities provided that the medium is observed at sufficiently high resolution.

Figure 5.7: Chaotic structures of typical mixtures (after Ottino, 1989).

Figure 5.8: Stretching between two points initially Δy apart after time δt. $v_x(y)$ is the horizontal velocity in (x,y).

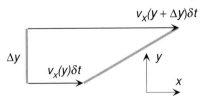

Where, in an agitated reservoir like the ocean or the mantle, are the effective mixing zones located? Let us consider (Fig. 5.8) a part of the reservoir where displacement is horizontal, velocity $v_x(y)$ is then dependent only on the distance y to the bottom. Two points located initially, one at (x, y) and the other on the same vertical at $(x, y + \Delta y)$, move after a time δt over a slightly different horizontal distance. This difference δl is written:

$$\delta l = v_x(y + \Delta y)\delta t - v_x(y)\delta t = \frac{dv_x}{dy}\Delta y \delta t \tag{5.19}$$

where the derivative of the velocity $v_x(y)$ is expressed in y. It can be seen that for a given initial vertical separation Δy, the term that governs the horizontal speed of stretching $\delta l/\delta t$ is the vertical variation dv_x/dy in horizontal speed, which is known in mathematical jargon as the velocity gradient. In general, this velocity gradient, which is a measure of the rate of deformation, and therefore of mixing, is particularly important in the turbulent zones of the ocean but also in boundary layers. In the mantle, complex geometries of the kind associated with mid-ocean ridges and subduction zones may also play an important role in reducing the size of heterogeneities.

Finally, when the thickness of the stretched and folded layers is reduced to very small length scales (typically 1–10 cm in the mantle), diffusion transport becomes an important smoothing process and small-scale heterogeneities disappear over time scales of 100–1000 Ma.

Because they are continuously recreated by the subduction of oceanic lithosphere, mantle heterogeneities are likely to be unevenly distributed even where stirring by convection is vigorous. The most probable chem-ical pattern in a randomly mixed medium is not chemical homogeneity. Let us illustrate this point through a naive comparison. A handful of rice thrown on a checkered table will never produce a regular distribution of grains over the surface. Instead, the grains will appear "randomly" spread (Fig. 5.9). By randomly, we mean that any small square is as likely to receive a rice grain as another. The number of grains one counts on any patch delimited on the table would be found nearly proportional to the surface area of the patch. One more time, we meet our old friend the Poisson distribution. As long as the mutual distance between chemically

Figure 5.9: One thousand grains of rice randomly thrown on a square table. The distribution is not regular, as each small square received a variable number of grains (2–20), but is perceived as random: a grain lands anywhere on the table with equal probability. The number of grains in any patch drawn on the table is proportional to the surface area of the patch (Poisson distribution). Likewise, the mantle is well mixed when the distribution of heterogeneities achieves maximum randomness. A regular distribution of heterogeneities is totally unlikely. Complex patterns of isotopic and chemical heterogeneities are no indication that the mantle is not well stirred.

or isotopically identifiable heterogeneities exceeds the mean diffusion length, vigorous mantle mixing will tend to produce a distribution of chemical and isotopic anomalies that resembles a Poisson distribution. Complex heterogeneity patterns in the mantle do not necessarily indicate imperfect mantle mixing but are a straightforward consequence of random redistribution of material by mantle convection.

References

Albarède, F. and Tamagnan, V. (1998) Modelling the recent evolution of the Piton de la Fournaise volcano, Réunion Island, 1931–1986. *J. Petrol.*, **29**, pp. 997–1030.

Lerman, A. (1988) *Geochemical Processes*. Malabar, Kreiger.

Garrels, R. M. and Mackenzie, F. T. (1971) Gregov's denudation of the continents. *Nature*, **231**, pp. 382–383.

Ottino, J. M. (1989) *The Kinematics of Mixing: Stretching, Chaos, and Transport.* Cambridge, Cambridge University Press.

Chapter 6
Waters present and past

The external aspects of geochemical cycles, the phenomena that occur at relatively low temperatures (typically from 0 to $+30\,°C$) in the ocean, the atmosphere, and in rivers, are largely governed by chemical equilibria in solution or at the water–mineral interface. The cycles themselves imply transfers controlled primarily by water–rock interaction (erosion, sedimentation, hydrothermal reactions) and by biological activity. A central role is played by the carbonate system. We will apply these concepts to the geochemistry of erosion and of the ocean, with a discussion of the impact of these cycles on climates in particular.

6.1 Basic concepts

A few important concepts that are part of college chemistry are required.

1. Acidity is the concentration (mol kg^{-1}) of [H^+] protons in a solution. The exact form, H^+ or H_3O^+, in which these protons occur is of little significance. A scale of acidity is defined by the potential pH of the protons in the solution, such that pH $= -\log$ [H^+]. At $25\,°C$, pure water has a pH of 7. A lower pH indicates an acidic solution and a higher pH a basic solution.
2. Ion behavior is dictated by the dissociation of acids and bases. An acid is a proton–donor molecule. A strong acid such as HCl or a strong base such as NaOH become completely dissociated to produce Cl^- and Na^+ ions, which behave essentially like inert species and are of relevance only in terms of charge. Weaker acids become partly dissociated by releasing one, two, or possibly more protons. In this way, carbonic acid H_2CO_3 dissociates into a bicarbonate ion HCO_3^- and then into a carbonate ion CO_3^{2-} by losing first

one and then two protons. The extent of dissociation is given by the mass action law:

$$\frac{[HCO_3^-][H^+]}{[H_2CO_3]} = K_1 \tag{6.1}$$

$$\frac{[CO_3^{2-}][H^+]}{[HCO_3^-]} = K_2 \tag{6.2}$$

Here the constants of the first (K_1) and second (K_2) dissociation vary with temperature and, to a lesser extent, with salinity. The notation $pK = -\log K$ is commonly used. A special case of dissociation is that of water by the reaction:

$$H_2O \Leftrightarrow H^+ + OH^- \tag{6.3}$$

for which the dissociation constant is:

$$\frac{[H^+][OH^-]}{[H_2O]} = K_{H_2O} \tag{6.4}$$

In the case where the solution is dilute, the $[H_2O]$ concentration of water in the solution is essentially constant and this relation can be rewritten in its more familiar form:

$$[H^+][OH^-] = K_w = 10^{-14} \tag{6.5}$$

3. Ion complexation is the association of ions carrying charges of opposite sign. In the ocean, for instance, the copper ion Cu^{2+} may be surrounded by different anions OH^-, Cl^-, HCO_3^-, which form different species of copper, but also by humic acids from the soil. In this sense, H_2CO_3 and HCO_3^- may be viewed as carbonate complexes of the proton. Complexation also obeys the mass action law with its successive constants.

4. Redox reactions relate to electron exchange. A reducing agent gives up electrons to an oxidant. Oxidation of Fe^{2+} into Fe^{3+} is a common result of electron acceptance by oxygen atoms. Oxidation of organic carbon (coal, petroleum) by atmospheric oxygen is among the most commonly employed of artificial energy-producing processes, but it also occurs naturally when sediments with a high detrital carbon content are eroded:

$$\begin{array}{ccc} C & + O_2 \Leftrightarrow & CO_2 \\ \text{sediment} & \text{air} & \text{atmosphere} \end{array} \tag{6.6}$$

Although it is possible to define an electron potential $p\varepsilon$ analogous to proton potential pH, the practice remains that of giving the measure by a potential E_h in volts, measured relative to a reference electrode. $p\varepsilon$ and E_h are proportional.

5. Atmospheric gases are scarcely soluble in water, except for CO_2. The solubility of gases is described by Henry's law, which establishes a simple proportionality between the partial pressure P_i of the gas i above the solution and its concentration in the solution. For CO_2, we write:

$$[H_2CO_3] = k_{CO_2}P_{CO_2} \tag{6.7}$$

where k_{CO_2} is the coefficient of solubility. The concentration of the H_2CO_3 species in natural waters is very low and the species actually present is dissolved CO_2, but for all practical purposes, the present notation is sufficient.

6. The solubility of solids precipitating from solutions is expressed using another coefficient of the mass action law, the product of solubility K_s. For carbonate precipitation $Ca^{2+} + CO_3^{2-} \Leftrightarrow CaCO_3$, the saturation condition is written:

$$\left[Ca^{2+}\right]\left[CO_3^{2-}\right] = K_s \tag{6.8}$$

7. The condition of electrical neutrality. This condition is normally written by calculating the charge balance of fully dissociated species Alk, which is known as alkalinity (not to be confused with basicity, which characterizes a solution with pH > 7):

$$Alk = \left[Na^+\right] + \left[K^+\right] + 2\left[Ca^{2+}\right] + 2\left[Mg^{2+}\right] + \cdots$$
$$- \left[Cl^-\right] - \left[NO_3^-\right] - 2\left[SO_4^{2-}\right] - \cdots \tag{6.9}$$

But since the neutrality of the solution must be maintained, we can also write:

$$Alk \approx \left[OH^-\right] - \left[H^+\right] + \left[HCO_3^-\right] + 2\left[CO_3^{2-}\right] \tag{6.10}$$

in which the borates and phosphates have been omitted. Alkalinity being a concentration of electrical charges, it is expressed in equivalent per kilogram (eq kg^{-1}). This is a measure of the neutralizing power of a solution: by adding HCl to a highly alkaline solution, the solution bubbles intensely as the carbonate ions are driven off and replaced by Cl^- ions. Notice that while the pH of a solution can vary with temperature or CO_2 pressure, alkalinity does not change and for this reason it is said to be "conservative." Alkalinity is conservative like [Na^+] or [Cl^-] but unlike [HCO_3^-] or [OH^-]. The sum [OH^-] − [H^+], known as caustic alkalinity, is normally negligible compared with carbonate alkalinity [HCO_3^-] + 2[CO_3^{2-}].

8. All the thermodynamic constants depend on temperature and, to a lesser extent, on pressure. Those analogous to an equilibrium coefficient may be expressed by a similar law to (2.11).

6.2 Speciation in solutions

An important problem in the geochemistry of solutions is the distribution of the chemical components among the various species. A simple example is that of the distribution of carbonates as H_2CO_3, HCO_3^-, and CO_3^{2-}. Calculating the speciation of a solution of known composition involves evaluating the abundance of different chemical species under prescribed conditions of temperature, pressure, pH, and any other factors. Let us make a very simple simulation of the deep ocean isolated from the atmosphere after the formation of bottom waters and which receives carbonate tests of sedimented foraminifera. We introduce m_{CO_2}

moles of CO_2 per liter into pure water, and then a very small known quantity of calcite, such that the molarity of calcite in water is m_{CaCO_3}, and leave the calcite to dissolve. Let us now inquire into the abundance of the species present in the solution. The CO_2 acidifies the water allowing the calcite to be dissolved.

We choose two components, Ca^{2+} and CO_2. The six possible species are OH^-, H^+, Ca^{2+}, and the carbonates H_2CO_3, HCO_3^-, and CO_3^{2-}. The concentration of H_2O is high enough compared with the rest of the dissolved species to be invariant in the process, and so we can eliminate it both as a species and a component. If we wish to determine the abundances of the six species, we have two equilibrium equations, one for the sum of CO_2:

$$[H_2CO_3] + [HCO_3^-] + [CO_3^{2-}] = m_{CO_2} + m_{CaCO_3} \qquad (6.11)$$

and the other for the sum of calcium ions:

$$[Ca^{2+}] = m_{CaCO_3} \qquad (6.12)$$

and one equation for electrical neutrality:

$$[HCO_3^-] + 2[CO_3^{2-}] + [OH^-] = [H^+] + 2[Ca^{2+}] \qquad (6.13)$$

We therefore need another three equations (given in the form of mass action laws) to determine the six unknowns. We choose, of course, (6.1) and (6.2) for carbonic acid dissociation, plus (6.5) for water dissociation.

This seemingly natural choice is in fact arbitrary. There is nothing to prevent us from replacing one of these equations by a combination of them. Thus, either of the two equations for carbonic acid dissociation could be replaced by the product of the two:

$$\frac{[CO_3^{2-}][H^+]^2}{[H_2CO_3]} = K_1 K_2 \qquad (6.14)$$

It can be seen that, in the general case, speciation calculations involve resolving systems that include non-linear equations and therefore call on advanced numerical techniques. There are many software packages that perform these calculations with a high degree of sophistication allowing for many complexes and the presence of solid or gaseous phases. Such software is economically significant for the exploration and management of water, oil, and geothermal resources. It can also be used for understanding weathering and the genesis of mineral deposits.

6.3 Water–rock reactions

Water has the effect on the minerals of igneous or metamorphic rocks of putting the more soluble cations into solution and producing residual clay minerals. A notable exception is quartz. A first type of reaction controlling interaction between the lithosphere and hydrosphere essentially involves reactions of proton and cation exchange such as:

$$2NaAlSi_3O_8 + \quad 2H^+ \quad + H_2O \Leftrightarrow Al_2Si_2O_5(OH)_4 + 4SiO_2 + \quad 2Na^+$$

(albite)   (solution)   (kaolinite)   (silica)   (solution)

$$(6.15)$$

Since albite is a very common constituent of crustal rocks, this reaction can also be seen as limiting the solubility of kaolinite, one of the most important clay minerals in soils. In the absence of feldspar, kaolinite solubility is controlled by a different proton-exchange reaction:

$$\tfrac{1}{2}Al_2Si_2O_5(OH)_4 + \quad \tfrac{7}{2}H_2O \quad \Leftrightarrow \quad H_4SiO_4 \quad + Al(OH)_4^- + \quad H^+$$

(kaolinite)   (solution)   (solution)   (solution)   (solution)

$$(6.16)$$

The complexation of ions in solution makes the calculation of mineral solubility more complicated. Ferric iron hydroxide $Fe(OH)_3$, another important mineral of soils, known as goethite, is a good example. Ferric iron speciation in near-neutral fresh water (chlorine would produce strong complexes) is controlled by the species Fe^{3+}, $Fe(OH)_2^+$, and $Fe(OH)_4^-$. The concentration of the dissolved species $Fe(OH)_3$ is essentially zero. We can write the following reactions of goethite dissolution:

$$Fe(OH)_3(s) + 3H^+ \Leftrightarrow Fe^{3+} + 3H_2O \qquad (6.17)$$

$$Fe(OH)_3(s) + H^+ \Leftrightarrow Fe(OH)_2^+ + H_2O \qquad (6.18)$$

$$Fe(OH)_3(s) + H_2O \Leftrightarrow Fe(OH)_4^- + H^+ \qquad (6.19)$$

where $Fe(OH)_3(s)$ stands for solid goethite and the other species are in solution. Using the equilibrium constants suitable for these two equations and considering that the concentration of a pure solid phase is unity, we obtain:

$$\frac{\left[Fe^{3+}\right]}{\left[H^+\right]^3} = 10^{3.2} \qquad (6.20)$$

$$\frac{\left[Fe(OH)_2^+\right]}{\left[H^+\right]} = 10^{-2.5} \qquad (6.21)$$

$$\left[Fe(OH)_4^-\right]\left[H^+\right] = 10^{-18.4} \qquad (6.22)$$

which enables us to calculate the total ferric content ΣFe^{3+} as:

$$\Sigma Fe^{3+} = \left[Fe^{3+}\right] + \left[Fe(OH)_2^+\right] + \left[Fe(OH)_4^-\right]$$

$$= 1.6 \times 10^3 \left[H^+\right]^3 + 3.2 \times 10^{-3} \left[H^+\right] + \frac{4.0 \times 10^{-19}}{\left[H^+\right]}$$

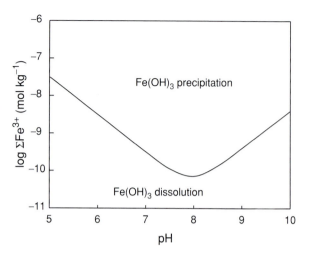

Figure 6.1: Ferric iron hydroxide (goethite) saturation in pure water as a function of pH. The minimum in the solubility curve results from the dominance of different hydroxide complexes at different pH values.

The results are plotted in Fig. 6.1. Ferric iron solubility is extremely small and, because of variable complexation by the different hydroxides, ΣFe^{3+} goes through a minimum for pH $\approx$ 8.

In other reactions, CO_2 is consumed by silicate dissolution, such as:

$$Mg_2Si_2O_6 + \quad 4CO_2 + \quad 2H_2O \Leftrightarrow 2SiO_2 + 4HCO_3^- + 2Mg^{2+}$$
$$\text{(pyroxene) (atmosphere)} \qquad \text{(silica)} \quad \text{(solution)} \quad \text{(solution)} \tag{6.23}$$

The CO_2 consumption reactions are not fundamentally different from the proton exchange reactions, but show the overlapping relationship between erosion, the carbon cycle – particularly marine carbonates and the CO_2 pressure of the atmosphere, and therefore, by way of the greenhouse effect, of climate. Proton-exchange reactions are very sensitive to the pH of solutions, since, allowing for a unit concentration of solids, the mass action law, for example for (6.15), requires a constant $[Na^+]/[H^+]$ ratio in the solution. Both acid rain, charged with sulfuric acid by oxidation of sulfur compounds at altitude, and ground water, charged with humic acids, will attack rock since their protons will be able to displace the Na^+, Ca^{2+}, or other ions of the minerals. Complexing anions (OH^-, Cl^-, CO_3^{2-}, SO_4^{2-}, PO_4^{3-}, humic and fulvic acids) also play a critical role in mineral solubility and transport. These reactions are greatly dependent on temperature. In contrast to limestone dissolution that recycles the fossil alkalinity produced since the origin of the Earth, the chemical erosion of silicates (Eqs. (6.15) and (6.23)) creates fresh alkalinity. CO_2 outgassed from the mantle will eventually be stored in newly formed sedimentary carbonates.

6.4 Biological activity

Biological activity has conditioned the chemistry of the ocean and the atmosphere since earliest times. It is characterized by exploitation of the environment to manufacture soft and hard body parts, and to maintain the metabolism of organisms. The primary plant producers extract carbon from the surrounding medium, atmospheric CO_2 in the case of land plants, and carbonates dissolved in seawater in the case of algae, and reduce it with solar energy (photosynthesis). The higher elements of the prey–predator chain, the food chain, recycle the primary reduced carbon. Organisms run on real reduced-carbon batteries. Respiration provides the oxygen for combustion and eliminates the CO_2 produced. While the reduced carbon, in the form of carbohydrates and lipids, acts as a repository of energy, it is also used to build soft parts, notably proteins. In addition to carbon, the production of organic matter requires nitrogen, phosphorus, and trace elements (particularly Fe, Mg, Zn), which are essential for manufacturing many vital enzymes.

For locomotion or defense, organisms form hard parts, based essentially on silica SiO_2 for diatoms, calcite $CaCO_3$ for foraminifera and invertebrates, and on phosphate for vertebrates. Carbonate precipitation consumes alkalinity. In the seas, calcite and silica are the main biogenic components of sediments and stand in contrast to the detrital material carried by rivers and the wind, and which is composed largely of clays, hydroxides, and quartz.

6.5 The carbonate system

By using methods from solution chemistry, it is possible to calculate the speciation of carbonates in water, for example the $\alpha_{HCO_3^-}$ fraction of carbonate dissolved as bicarbonate ions. This parameter is defined as:

$$\alpha_{HCO_3^-} = \frac{[HCO_3^-]}{[CO_3^{2-}] + [HCO_3^-] + [H_2CO_3]} \tag{6.24}$$

Dividing this equation by $[H_2CO_3]$, utilizing the two dissociation constants of (6.2), and multiplying by $[H^+]^2$ gives:

$$\alpha_{HCO_3^-} = \frac{K_1[H^+]}{K_1 K_2 + K_1[H^+] + [H^+]^2} \tag{6.25}$$

with similar expressions for $\alpha_{CO_3^{2-}}$ and $\alpha_{H_2CO_3}$. If these expressions are plotted against pH, three dominance zones can be identified, separated by the pK_1 (6.4) and pK_2 (10.3) of carbonic acid (Fig. 6.2). At low pH values, the surplus of protons leaves all the carbonates as carbonic acid H_2CO_3, while at high pH the deficit of protons favors CO_3^{2-}. HCO_3^- is the dominant ion in the intermediate pH zone. Natural water is weakly acidic

Figure 6.2: Speciation of
dissolved carbonates as a
function of the pH of the
solution (see (6.25)). Notice
that the predominance of
each carbonate species
changes when pH values
equal to pK are crossed.
Much natural water has a pH
close to neutral and is
therefore dominated by the
bicarbonate ion HCO$_3^-$.

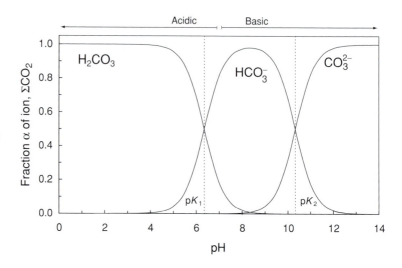

Figure 6.2: Speciation of dissolved carbonates as a function of the pH of the solution (see (6.25)). Notice that the predominance of each carbonate species changes when pH values equal to pK are crossed. Much natural water has a pH close to neutral and is therefore dominated by the bicarbonate ion HCO$_3^-$.

or weakly basic. Seawater, with a pH of 7.6–8.0, is largely dominated by HCO$_3^-$ ions.

The carbonate system in a solution in contact with the atmosphere is controlled by the chemical variables P_{CO_2}, pH, alkalinity $Alk \approx \left[\mathrm{HCO_3^-}\right] + 2\left[\mathrm{CO_3^{2-}}\right]$, the sum of available carbonates $\Sigma CO_2 \approx \left[\mathrm{HCO_3^-}\right] + \left[\mathrm{CO_3^{2-}}\right]$ (omitting H$_2$CO$_3$ and atmospheric CO$_2$), and the physical variables of temperature and, to a lesser extent, pressure. These variables are not independent and we seek to work with conservative variables, which exclude pH and carbonate concentrations. Combining the expressions for Alk and ΣCO_2 gives:

$$\left[\mathrm{CO_3^{2-}}\right] = Alk - \Sigma CO_2$$
$$\left[\mathrm{HCO_3^-}\right] = 2\Sigma CO_2 - Alk \tag{6.26}$$

In seawater (Fig. 6.3), alkalinity varies from 2.0 to 2.4 meq kg^{-1} (the equivalent is a number of charges expressed in molar units). Some 90% of dissolved carbonate is in the form [HCO$_3^-$]. Alkalinity is much lower, or even negative, in rain water and river water. By combining these expressions with those describing the solubility of CO$_2$ and the dissociation of carbonic acid, we obtain the very important equation:

$$P_{CO_2} = \frac{K_2}{k_{CO_2}K_1} \frac{(2\Sigma CO_2 - Alk)^2}{(Alk - \Sigma CO_2)} \tag{6.27}$$

This expression is plotted for the ocean–atmosphere systems in Fig. 6.4, where CO$_2$ pressures are expressed in ppmv (parts per million volume) and it is taken that at 25 °C, k_{CO_2} is approximately equal to 0.029 moles per kg and per atm. The pH of seawater can be calculated either in surface water at a known pressure P_{CO_2}, or in deep water for a

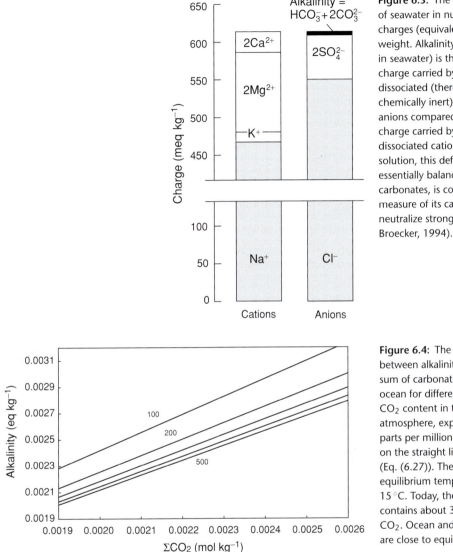

Figure 6.3: The composition of seawater in number of charges (equivalent) per unit weight. Alkalinity (2 meq kg^{-1} in seawater) is the deficit of charge carried by totally dissociated (therefore chemically inert) anions compared with the charge carried by totally dissociated cations. In a solution, this deficit, which is essentially balanced by carbonates, is constant. It is a measure of its capacity to neutralize strong acids (after Broecker, 1994).

Figure 6.4: The relationship between alkalinity and the sum of carbonates in the ocean for different values of CO_2 content in the atmosphere, expressed in parts per million volume on the straight lines (Eq. (6.27)). The assumed equilibrium temperature is 15 °C. Today, the atmosphere contains about 350 ppmv CO_2. Ocean and atmosphere are close to equilibrium.

fixed value of ΣCO_2, by introducing the values of $\left[HCO_3^-\right]$ and $\left[CO_3^{2-}\right]$ obtained through (6.26) into the dissociation (6.2).

Let us consider a different example relevant to our oceanic system in which the calcium concentration of the ocean is fixed by carbonate saturation and seawater maintained at a constant atmospheric pressure of CO_2. Equations (6.11) and (6.12) have to be replaced by two equations expressing saturation in calcite:

$$\left[Ca^{2+}\right]\left[CO_3^{2-}\right] = K_s \qquad (6.28)$$

and CO_2 solubility:

$$[H_2CO_3] = k_{CO_2} P_{CO_2} \qquad (6.29)$$

The total concentration ΣCO_2 of carbonate ions becomes:

$$\Sigma CO_2 = \frac{K_s}{[Ca^{2+}]}\left(1 + \frac{[H^+]}{K_2} + \frac{[H^+]^2}{K_1 K_2}\right) \qquad (6.30)$$

$$P_{CO_2} = \frac{K_s}{K_1 K_2 k_{CO_2}}\frac{[H^+]^2}{[Ca^{2+}]} \qquad (6.31)$$

As in the two cases considered before, (5.12)–(5.19), there are two degrees of freedom of the carbonate system introduced by the interplay of geological conditions. Pairs of parameters, such as alkalinity and ΣCO_2, pH and P_{CO_2}, calcite saturation and P_{CO_2}, can be imposed, and all of the other variables follow from them.

Reconstructing the composition of the ancient oceans therefore comes down to finding geochemical tracers (proxies) for evaluating one or other of these parameters. Much hope is pinned on the capacity of isotopic concentrations of boron measured from the calcareous tests of foraminifera to reflect the pH of the early oceans. The boric acid of seawater is a weak acid and boron is found in two different states governed by the reaction:

$$H_3BO_3 + H_2O \Leftrightarrow B(OH)_4^- + H^+ \qquad (6.32)$$

We note the apparent dissociation constant of H_3BO_3 as K_B ($pK_B = 8.75$) and the product of the equilibrium constant of this reaction by the $[H_2O]$ molarity of water, which is very close to 55.6 in all natural waters. The proportion of the two species clearly depends on the pH of the solution and we can formulate the fraction of boron stored for each of them, for example $\varphi_{H_3BO_3}$, in the usual way:

$$\varphi_{H_3BO_3} = \frac{[H_3BO_3]}{[H_3BO_3] + [B(OH)_4^-]} = \frac{[H^+]}{[H^+] + K_B} \qquad (6.33)$$

Clearly, the sum of $\varphi_{H_3BO_3}$ and $\varphi_{B(OH)_4^-}$ is unity. Moreover, an isotopic equilibrium equation between the two boron carrier species can be written:

$$\frac{(^{11}B/^{10}B)_{H_3BO_3}}{(^{11}B/^{10}B)_{B(OH)_4^-}} = \alpha^B(T) \qquad (6.34)$$

where $\alpha^B(T)$ is the isotopic fractionation coefficient between the two species of boron present in the solution. The residence time of boron in seawater being several million years, it can be considered that neither its content nor its overall isotopic composition change between glacial and

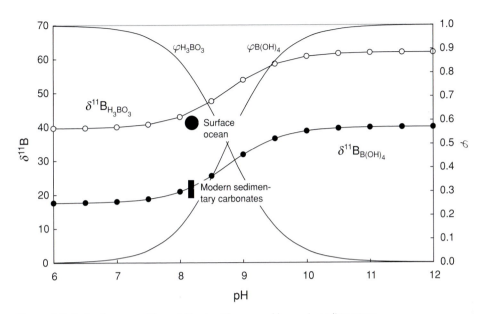

Figure 6.5: Isotopic composition of dissolved boron and boron in sedimentary carbonates as a function of the pH of seawater. The $\delta^{11}B$ values are defined with reference to a standard boric acid (SRM 951). Fraction φ and $\delta^{11}B$ of each species H_3BO_3 and $B(OH)_4^-$. There is an isotopic fractionation of ^{11}B of 22 per mil between the two species. The $\delta^{11}B$ of $B(OH)_4^-$ incorporated in the sedimentary carbonate is a measure of the pH of the ocean from which limestone forms.

interglacial periods, which alternate over a time scale of 100 000 years. The mass balance condition for the $^{11}B/^{10}B$ ratio can therefore be written as:

$$\left(\frac{^{11}B}{^{10}B}\right)_{ocean} = \varphi_{H_3BO_3}\left(\frac{^{11}B}{^{10}B}\right)_{H_3BO_3} + \varphi_{B(OH)_4^-}\left(\frac{^{11}B}{^{10}B}\right)_{B(OH)_4^-} \quad (6.35)$$

Only the charged borate $B(OH)_4^-$ seems to be incorporated in the calcite of foraminifera. The previous two equations can therefore be combined in the form:

$$\left(\frac{^{11}B}{^{10}B}\right)_{foram} = \left(\frac{^{11}B}{^{10}B}\right)_{B(OH)_4^-} = \frac{\left(^{11}B/^{10}B\right)_{ocean}}{\varphi_{H_3BO_3}\alpha^B(T) + \left(1 - \varphi_{H_3BO_3}\right)} \quad (6.36)$$

If we know the $^{11}B/^{10}B$ ratio of the ocean from measurement of present-day seawater, the growth temperature of foraminifera (given by oxygen isotopes), and isotopic fractionation $\alpha^B(T)$ (0.978 at 20 °C), we can deduce $\varphi_{H_3BO_3}$ and therefore, in principle, the pH of the oceans in which the foraminifera formed (Fig. 6.5). This method assumes that we have identified species for which isotopic fractionation of boron occurs reproducibly, which is a significant limitation for the time being.

Figure 6.6: The atmospheric water cycle and the δ^{18}O (in per mil) of various types of precipitation. Vapor formed by evaporation of the warm surface ocean between the tropics migrates with the atmospheric circulation towards the poles. Incremental precipitation in rain at low latitude and in snow at high latitude progressively depletes atmospheric water vapor, rain, and snow in ^{18}O. A similar poleward depletion of the atmosphere takes place for deuterium with respect to hydrogen, thereby producing the relationship of Fig. 6.7.

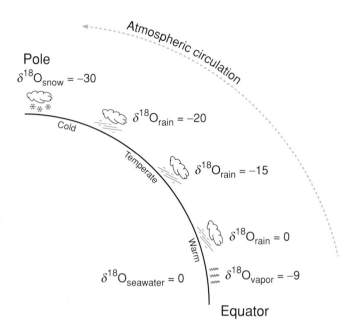

6.6 Precipitation, rivers, weathering, and erosion

The hydrological cycle is well understood (Fig. 6.6). Evaporation caused by intense solar heating of the oceans between the tropics produces water vapor. When the hot, moist air rises, the water vapor condenses into fine droplets forming clouds, but only falls to the ground as rain when the droplets coalesce to attain sufficient size. The masses of hot air migrate toward the poles and as they cool precipitation progressively drains the moisture from the atmosphere. This can be seen from the measurement of isotopic concentrations of hydrogen and oxygen in precipitation, i.e. rain (meteoric) water and snow (Fig. 6.7). Because heavy isotopes concentrate preferentially in liquid water and ice, progressive condensation of atmospheric vapor produces a very clear poleward depletion of precipitation in the heavier isotopes of hydrogen and oxygen. A correlation ensues between δD (a measure of the ^{2}H/^{1}H ratio) and δ^{18}O of rain or snow, corresponding to this progressive depletion of deuterium (D or ^{2}H) and ^{18}O in accordance with the linear approximation of fractional condensation:

$$\frac{d\delta D}{d\delta^{18}O} = \frac{\alpha^{D}(T) - 1}{\alpha^{18O}(T) - 1} \tag{6.37}$$

where coefficients α measure the given isotope fractionation between liquid and vapor at temperature T of the clouds. Strictly, this expression should be corrected to allow for kinetic effects during phase changes. This expression yields the relationship $\delta D = 8\,\delta^{18}O + 10$ (Fig. 6.7). Rain

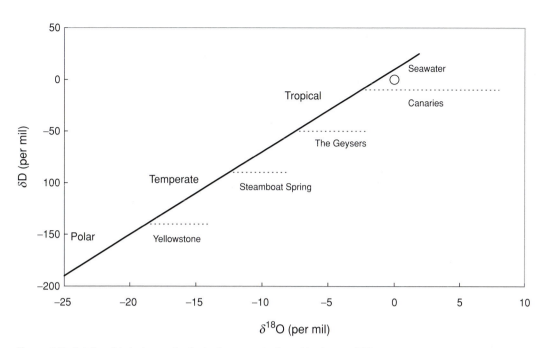

Figure 6.7: Relationship between the isotopic concentration of hydrogen (δD) and oxygen ($\delta^{18}O$) in meteoric waters (solid line). Liquid precipitation depletes the atmospheric water vapor in heavy isotopes as the clouds move poleward. Most ground waters plot along the meteoric line ($\delta D = 8\delta^{18}O + 10$) demonstrating their meteoric origin and negligible interaction with minerals in the first kilometer of the crust. Many geothermal springs (dotted lines) have the same δD as local rain water. As there is practically no hydrogen in the crust, this indicates their meteoric origin, with the $\delta^{18}O$ variations reflecting exchange with crustal mineral oxygen at temperatures of about 75–350 °C.

waters from temperate regions and polar ice are depleted with respect to seawater in the light isotope by up to several percent for oxygen and up to tens of percent for hydrogen. This relationship does not pass through the composition of seawater because evaporation takes place in warm surface water while condensation in clouds takes place at much lower temperatures. A somewhat similar distribution is found with altitude: as air rises in mountain areas, the various types of precipitation become depleted in D and ^{18}O.

This relationship requires that almost all underground water is re-cycled rain water. Geothermal springs lie on a horizontal line through the isotopic composition of the rainfall of the location. This indicates that the ground water is rain water in which oxygen has exchanged its isotopes with the surrounding rock; as the rock is virtually devoid of hydrogen, δD is left unchanged. Therefore, there is no such a thing as "juvenile" spring water, i.e. water that comes from the depths of the Earth without ever having seen the surface. The same is true of hydrothermal springs.

The mineral contribution of sea spray to rain water smoothly decreases with distance from the coast. This gradient involves Na^+ and Cl^- ions above all, the readily identifiable so-called cyclical ions. The remainder of the water infiltrates or runs off, contributing to erosion.

Chemical erosion (weathering) can be formulated as a set of dissolution reactions. Silicates formed at high temperature in igneous or metamorphic environments (olivine, pyroxenes, feldspars, micas) are metastable in the low-temperature conditions of the surface. They react with precipitation to form clay minerals, that are the stable form of silicates, hydroxides of the most insoluble ions in their oxidized form (Fe^{3+}, Al^{3+}), while the most soluble ions (Na^+, K^+, Ca^{2+}, Mg^{2+}) are released in the runoff and river water. Silica is scarcely soluble and quartz from any origin is largely left untouched by erosion, but the level of $[SiO_2]$ in rivers is comparable to the concentration of other major elements.

It is kinetic barriers that preserve high-temperature mineral assemblages such as those of granite and gneiss over very long geological times at the Earth's surface, whereas equilibrium thermodynamics requires that feldspars and many other silicates should have turned to clay. Erosion studies must therefore pay due attention to dissolution rates. Most erosion processes are dissolution/reprecipitation reactions resembling the albite destruction described by (6.15). Normally, dissolution is the rate-limiting step, and the nature of dissolved cations is not critical in controlling the reaction rate, with the notable exception of $[H^+]$. Quite commonly, rainwater pH values fall in a range of 4–6. Acid waters have the greatest erosion potential because abundant $[H^+]$ displaces the major soluble ions (Na^+, K^+, Ca^{2+}, Mg^{2+}) constituting the rock-forming minerals. Once normalized to surface area and after establishment of steady state, most mineral dissolution rates R may adequately be described by the equation:

$$\ln R = \ln R_0 + n\, \mathrm{pH} \tag{6.38}$$

where R_0 and n are constant for a specified mineral. Temperature dependence of the dissolution rate is very critical: warm climates promote chemical erosion. This dependence takes the usual form:

$$\ln R = A - \frac{E}{RT} \tag{6.39}$$

where A is a constant and E the activation energy of this particular reaction. Values of n between 0 and -1 are common, while E usually falls between the activation energy of ionic diffusion in solution and the energy required to break the silicate bond. Weathering rates typically increase by

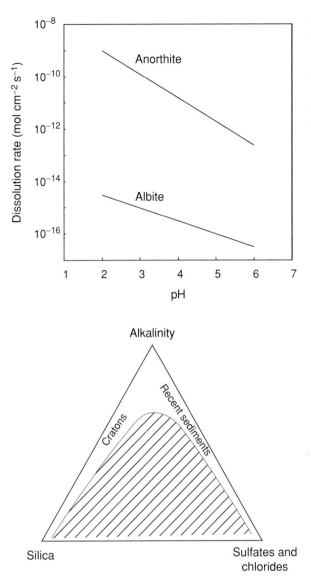

Figure 6.8: Dependence of albite and anorthite weathering rates on pH. Acid waters dissolve feldspars faster than neutral waters. These rates typically increase by a factor of ten when temperature increases by 25 °C.

Figure 6.9: Compositional distribution of the dissolved charge of rivers among its three main components: silica, alkalinity (mainly carbonates), and chlorides and sulfates derived mostly from evaporites. These compositions are closely controlled by the nature of the substrate (granitic cratons, sedimentary platform). The hatched zone contains no data (after Stallard and Edmond, 1983).

a factor of ten when temperature increases by 25 °C. The pH-dependence property of the weathering rates is illustrated for feldspars in Fig. 6.8.

The major features of river chemistry may be summarized with a small number of parameters. After subtraction of cyclic ions, the remainder of the dissolved mineralization of fresh water can be divided into carbonate alkalinity (from dissolution of silicates by the reactions described above or from redissolution of limestone), sulfate and chloride from dissolution of evaporites, and silica from dissolution of silicates (Fig. 6.9). These quantities vary greatly with the site where water erodes the soil. In addition to dissolved mineralization, rivers transport abundant mineral (clay and iron hydroxides) and organic colloids.

6.7 Elements of marine chemistry

Rivers carry a dissolved mineral load to the ocean. It is practical to group all of this added mineralization under the heading of erosion-related alkalinity flux. Many elements do not pass the filter of estuaries; at this point, fresh water and salt water mix, causing a substantial increase in the dissolved charges (ionic strength) in water. The abundant hydrous ions polarize thereby reducing the electrical field within the solution and mutual repulsion between charged particles. Massive precipitation of colloids takes place, that adsorb many elements such as the transition elements, rare-earths, etc. This phenomenon is amplified by the richness of estuaries in organic matter.

The ocean surface is characterized by abundant life, sustained throughout the depth of water where light can penetrate (known as the photic zone, some 50 m deep) by primary photosynthetic production by algae. This primary production allows other planktonic forms or larger organisms to develop by grazing and predation. As indicated above, organic matter concentrates light isotopes, particularly ^{12}C relative to ^{13}C and ^{14}N relative to ^{15}N, and the intensity of this productivity can be measured with the increased $\delta^{13}C$ and $\delta^{15}N$ of surface water compared with deep water (Figs. 6.10 and 6.11). Biological activity consumes the surplus alkalinity supplied by rivers, notably in the form of $CaCO_3$, which is one of the major constituents of the hard parts of the microfauna (foraminifera). Biological activity is limited only by the availability of other elements required for tissue construction (P, N), for metabolism (Fe), and for locomotion and defense (SiO_2). It is a remarkable fact that oceanic organisms, including phyto- and zooplankton, consume these constituents in fairly constant proportions, particularly the C : N : P proportions which are universally 115 : 15 : 1 in organic matter (Redfield ratio) . These proportions are believed to correspond approximately to a fairly general reaction, the Redfield–Ketchum–Richards (RKR) reaction, which leads to the production of biological material:

$$106CO_2 + 16HNO_3 + H_3PO_4 + 122H_2O \Leftrightarrow$$
$$(CH_2O)_{106}(NH_3)_{16}H_3PO_4 + 138O_2$$

This reaction is endothermic and uses solar energy (photons) to produce carbon and nitrogen in a highly reduced state that later may liberate large amounts of chemical energy. The elements used by biological activity, such as C, N, P, and Si, are depleted in surface water and are known as nutrients. They contrast with the residual major elements, such as Na,

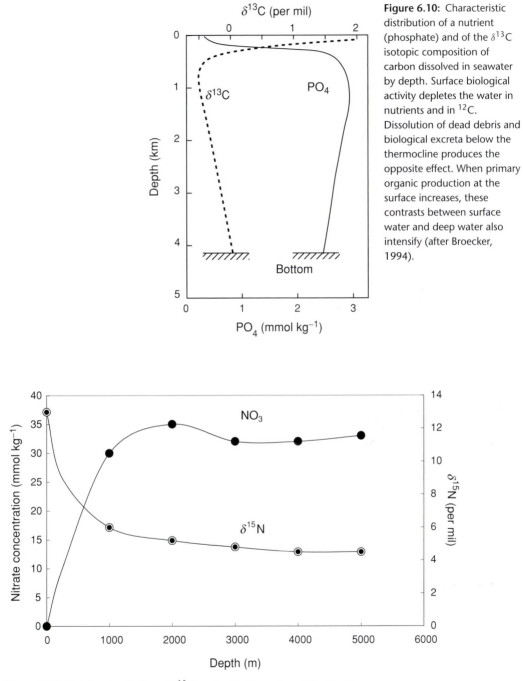

Figure 6.10: Characteristic distribution of a nutrient (phosphate) and of the $\delta^{13}C$ isotopic composition of carbon dissolved in seawater by depth. Surface biological activity depletes the water in nutrients and in ^{12}C. Dissolution of dead debris and biological excreta below the thermocline produces the opposite effect. When primary organic production at the surface increases, these contrasts between surface water and deep water also intensify (after Broecker, 1994).

Figure 6.11: Variation in nitrate and $\delta^{15}N$ content in the waters of the Southern Ocean: intense surface biological activity depletes the dissolved nitrate and, as with carbon (Fig. 6.10), preferentially removes the light isotope, here ^{14}N (after Sigman *et al.*, 1999).

Figure 6.12: Elements used
by biological activity, such as
C, N, P, Si (nutrients), are
depleted at the surface.
Residual major elements, such
as Na, Cl, K, Mg, are not
affected by biological activity.
Al is a detrital element
introduced into the
thermocline by airborne
particles. Dissolved oxygen is
equilibrated with atmospheric
gases near the surface but is
significantly depleted below
the thermocline as a result of
the oxidation of falling
organic debris.

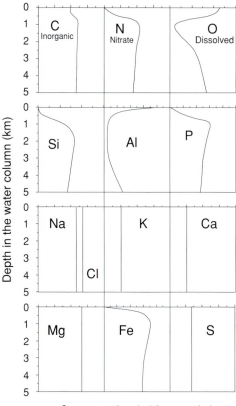

Figure 6.12: Elements used by biological activity, such as C, N, P, Si (nutrients), are depleted at the surface. Residual major elements, such as Na, Cl, K, Mg, are not affected by biological activity. Al is a detrital element introduced into the thermocline by airborne particles. Dissolved oxygen is equilibrated with atmospheric gases near the surface but is significantly depleted below the thermocline as a result of the oxidation of falling organic debris.

Cl, K, and Mg, which are not affected by biological activity (Fig. 6.12). In the section on chromatography, we saw that the reactivity of some ions (Cd, Ba, Fe, rare-earths, Th) relative to biological particulate matter rich in nitrate, phosphate, and silica accelerates their transfer from the surface to the ocean floor, even if they are not directly involved in the biological cycle. Such ions therefore display a nutrient-like character with depletion in the surface water and regeneration in deep water.

The waste matter and carrion produced by surface organisms sink through the ocean where they are recycled by other life forms. In the layer of water down to 100–1000 m, known as the thermocline because temperature varies very quickly, nitrogen, phosphorus, and other nutrients are re-dissolved. This process is accompanied, of course, by the consumption of oxygen (respiration), which is therefore depleted at such depths relative to the remainder of the ocean (oxygen minimum), and by the release of CO_2.

Some 95% of organic matter is recycled in this way before it sinks to great depths. Nutrients are recycled through the thermocline many

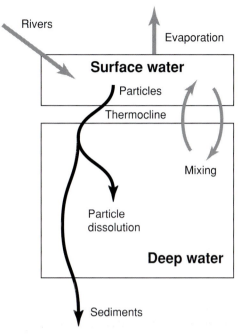

Figure 6.13: The two-reservoir model of the ocean of Broecker and Peng (1982).

times before being incorporated into the sediment. The importance of recycling has been demonstrated by Broecker and Peng (1982) in a simple but illustrative way (Fig. 6.13). Let us consider an ideal ocean made of two layers one on top of the other, surface water and deep water, separated by the thermocline. Elements are carried from the continents to the sea by rivers and are removed from the deep ocean by sedimentation. Let us call F_{river} the input of river water (kg m^{-3} s^{-1}) into the ocean and C_{river}^i the concentration (mol kg^{-1}) of river water in element i. F_{river} is compensated by evaporation so that the amount of surface water remains unchanged. We call F_{mix} the amount of water exchanged per unit time between the surface and deep-water layers, C_{deep}^i and $C_{surface}^i$ the concentration of element i in deep and surface water, respectively. P_{part}^i stands for the flux of i carried downward by sinking particles. We first write that, at steady state, inputs and outputs of i must be equal so that:

$$F_{river}C_{river}^i + F_{mix}C_{deep}^i = F_{mix}C_{surface}^i + P_{part}^i \qquad (6.40)$$

The fraction g of the downgoing flux of i carried by particles is therefore:

$$g = \frac{P_{part}^i}{F_{mix}C_{surface}^i + P_{part}^i} \qquad (6.41)$$

or:

$$g = 1 - \frac{\frac{F_{mix}}{F_{river}} \frac{C^i_{surface}}{C^i_{river}}}{1 + \frac{F_{mix}}{F_{river}} \frac{C^i_{deep}}{C^i_{river}}} \tag{6.42}$$

Because, at steady state, dissolved elements that enter the ocean in rivers must leave it as sediments, the fraction f of settling particles that eventually exit the deep-water reservoir as sediments is such that:

$$f P^i_{part} = F_{river} C^i_{river} \tag{6.43}$$

and therefore:

$$f = \frac{1}{1 + \frac{F_{mix}}{F_{river}} \left(\frac{C^i_{deep}}{C^i_{river}} - \frac{C^i_{surface}}{C^i_{river}} \right)} \tag{6.44}$$

The quantity $1 - fg$ indicates the proportion of the riverine input of i to the ocean that is recycled through the thermocline instead of being sedimented. Using the theory developed above (Eq. (5.8)) to evaluate F_{mix} we obtain $F_{mix}/F_{river} \approx 30$. For phosphates, we use the values $C^i_{deep}/C^i_{river} = 3$ and $C^i_{surface}/C^i_{river} = 0.15$, which emphasizes the depletion of surface waters in nutrients. We therefore obtain $g = 0.95$ (95% of the riverine input of phosphate is exported to the deep ocean as particles), $f = 0.01$, and $fg = 0.01$, which indicates that 1% of the riverine phosphate is actually exported to sediments while 99% is recycled through the thermocline by upwelling. For silica we would obtain 95%. An alternative statement is that a phosphorus atom is recycled 100 times and a silicon atom 20 times through the thermocline before they find their way into the sediments.

The downward transfer of nutrients is therefore achieved by settling of the particles and not by the sinking of surface waters depleted by biological activity. In contrast, upward transfer of nutrients is most efficiently effected by deep-water upwelling. This mechanism explains why deep-water upwellings off the coasts of Mauritania, Peru, Namibia, etc., are associated with intense primary productivity.

The recycling of soft parts is accompanied by the re-dissolution of calcite from carbonate tests. Alkalinity and ΣCO_2 increase while pH decreases with depth, typically from values of 8.2 in surface water to 7.8 in deep water. As discussed earlier, the "surface-water" and "deep-water" boxes may, by way of fractionation throughout the thermocline, maintain a sharp contrast in spite of some nutrients having very long overall residence times.

The deep ocean is cold and isolated from the atmosphere, contrary to surface water. It is undersaturated in $CaCO_3$ whereas surface water is saturated, explaining why living organisms have found an energy advantage in using this compound preferentially for the fabrication of hard parts. The saturation limit, or lysocline, and the "carbonate compensation depth" (CCD), which is the depth below which carbonates are not observed in sediments, currently lie at a depth of about 4500 m. Above this level, carbonate sediment is abundant, as on either side of the mid-ocean ridges, which may rise to just 2500 m below the surface; at greater depths, all calcite is dissolved and only residue is deposited, very slowly, as red clay, the composition of which is much affected by atmospheric dust.

In addition to vertical chemical variations, horizontal variations are important too. This is partly because the largest rivers (e.g. the Amazon, Mississippi, and Congo) flow into the Atlantic, which therefore receives a greater supply of nutrients than the Pacific, hence its rich biota. The largest mass of deep water forms in the North Atlantic by downwelling of the very salty water of the Gulf Stream, made even denser at high latitudes by the formation of the ice cap (Fig. 6.14). This is the North Atlantic Deep Water (NADW). Other masses form near the Antarctic, such as the Antarctic Bottom Water (AABW) in the Weddel Sea. This very cold water, which is less salty than the NADW as it is less subjected to evaporation, lies at the bottom of the main oceans. The NADW begins its long journey along the coast of the Americas to the Indian Ocean and the Pacific Ocean by way of the circum-Antarctic current (Fig. 6.15). There it rises to the surface and returns to the Atlantic via Indonesia and Madagascar. The whole oceanic circulation system is driven by temperature and salinity contrasts (thermohaline circulation) and is often referred to, for obvious reasons, as the oceanic "conveyor belt." In Chapter 5 we calculated the age for renewal of the deep water at 1600 years, which is approximately the duration of a cycle. During this deep journey, the water that forms at the surface, and so is poor in nutrients but rich in oxygen, receives surface organic input. As the debris raining from the surface is dissolved at depth, deep water progressively gains nutrients and alkalinity, and loses oxygen. It also gains CO_2 through the effects of respiration, which is reflected by the lower carbonate sediment production in the Pacific than in the Atlantic. Generally, these characteristics clearly distinguish the old water of the deep Pacific from the young Atlantic waters.

6.8 Climate

Atmospheric carbon dioxide is a greenhouse gas. When the Sun's rays strike the ground, they are reflected but at a longer wavelength, in the infrared. We can feel this radiation when we walk along a paved road at night after a sunny summer's day. This reflected radiation is absorbed by

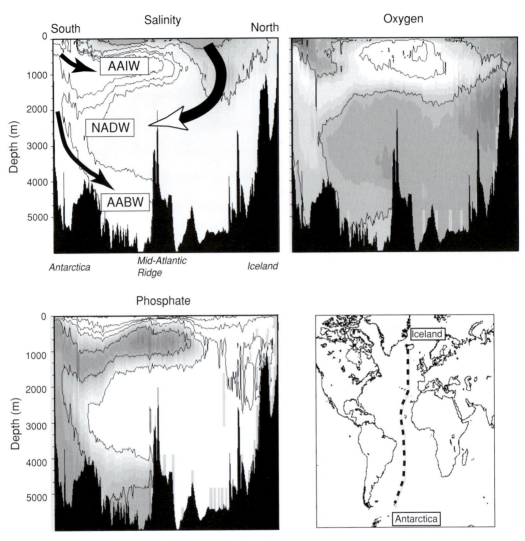

Figure 6.14: Formation and circulation of Atlantic water. The cross-sections follow the dashed line shown in the bottom right panel. The Gulf Stream, rich in salt because of surface evaporation in the intertropical zone, cools in winter at high latitudes and sinks to form the North Atlantic Deep Water (NADW). This new deep water moves southward at a depth of about 3500 m along the coast of America. It is salty and oxygenated but depleted in nutrients, such as phosphate, by the intense biological activity of the North Atlantic. The Antarctic Bottom Water (AABW) forms in the Weddel Sea. This very cold water is less subjected to evaporation and receives melt water from the Antarctic ice sheet: it is therefore less saline than NADW. It is also richer in nutrients, especially phosphate, as it recycles old water returned from the Pacific by the circum-Antarctic current. It is the most dense of all water masses and composes the bottom water of the main oceans. Intermediate Atlantic Water, diluted by rain in the southern mid-latitudes, is lighter. Figures drawn using OceanAtlas.

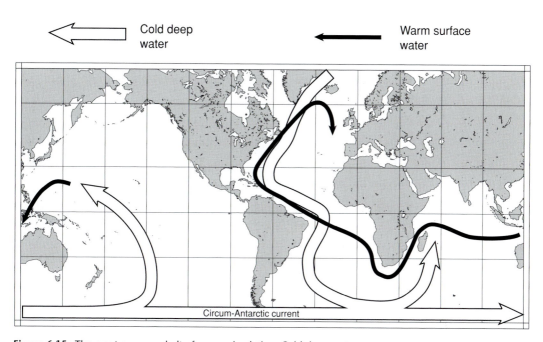

Cold deep water

Warm surface water

Circum-Antarctic current

Figure 6.15: The great conveyor belt of ocean circulation. Cold deep water forms in the North Atlantic by downwelling of the very salty water of the Gulf Stream and travels along the coast of the Americas. It then merges with the circum-Antarctic current, which feeds the Indian Ocean and the Pacific Ocean. The return flow of warm surface water from the Pacific passes through the Indonesia straits into the Indian Ocean, and circulates around the tip of South Africa into the Atlantic where it joins the Gulf Stream.

CO_2 and other greenhouse gases (especially water vapor, methane, and ozone) and heats the atmosphere. The history of the Earth's climate is therefore for a large part that of its atmospheric CO_2.

This effect can be seen clearly in the composition of Quaternary sediments where the extent of carbonate preservation varies with the depth of deposition in keeping with the pattern of glaciations (Fig. 6.16) that recur every $\approx 100\,000$ years. This pattern is similar to the pattern of fluctuations in marine oxygen isotopes as recorded by the $\delta^{18}O$ of the calcite of deep-water foraminifera. During glacial times, a greater quantity of ice with very negative $\delta^{18}O$ values is stored in the polar ice caps and seawater oxygen is enriched in ^{18}O (Fig. 6.17). As an effect of mass balance, the glacial $\delta^{18}O$ maxima in seawater and the carbonates in isotopic equilibrium with it (maximum ice volume) are mirrored by $\delta^{18}O$ minima in the ice: the preferential entrainment of heavy ^{18}O into rain and snow is more efficient during cold periods.

Lower average temperatures during the ice ages are explained by the astronomical modulation of insolation and its amplification by a lower

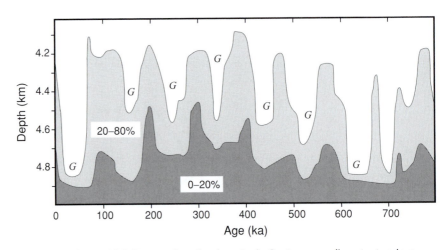

Figure 6.16: Preservation of carbonates in Quaternary sediments at various depths. The zone between the two curves mainly depicts the transition of the carbonate compensation depth (CCD). In glacial periods (*G*), the CO_2 content of the atmosphere is lower, the ocean less acidic, the CCD sinks, and the calcareous fraction of sediments increases (after Farrell and Prell, 1989).

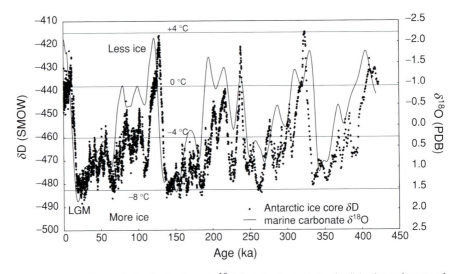

Figure 6.17: Fluctuation of $\delta^{18}O$ (relative to PDB standard) in the carbonate of benthic foraminifera (SPECMAP database). Increased storage of water in polar ice during glacial periods increases the ^{18}O isotope concentration of the ocean. The $\delta^{18}O$ curve follows (although with a small delay) the temperature curve deduced from the δD values measured in the Vostok ice core in Antarctica (Petit *et al.*, 1999). The horizontal lines show the temperature shifts in the polar region deduced from the δD values in ice. LGM = last glacial maximum. The periodic aspect of fluctuations is caused by astronomical forcing (Milankovic cycles).

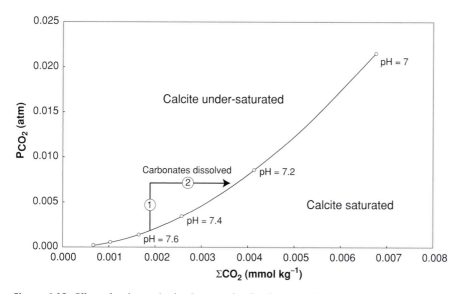

Figure 6.18: Effect of a change in the P_{CO_2} on the dissolution and precipitation of carbonates. The curve represents the pressure of carbon dioxide in the atmosphere above a solution saturated in calcite at different values of ΣCO_2. ① Sudden P_{CO_2} increase. ② Reaction of the excess dissolved CO_2 with carbonate sediments. An opposite effect (precipitation of carbonates) would be obtained for a drop in P_{CO_2}.

level of CO_2 in the atmosphere, and therefore a weaker greenhouse effect. Plotting P_{CO_2} vs ΣCO_2 for different pH values for seawater saturated in calcite as given by (6.30) and (6.31), we can see the effect of changing P_{CO_2} on the precipitation or dissolution of calcium carbonate (Fig. 6.18). A reduced P_{CO_2} decreases ΣCO_2, makes the ocean less acidic and therefore enhances carbonate preservation, which is reflected by a deepening of the CCD. The intervals during which limestone is abundant at great depths are cold periods.

The variability in the isotopic composition of oxygen recorded by foraminifera reveals two essential climatic phenomena:

1. The abundance of polar ice, which varies periodically with orbital forcing, especially the precession of the equinoxes (19 000 and 23 000 years), the variation in the inclination (obliquity) of the Earth's axis of rotation on the ecliptic (41 000 years), and the eccentricity of the Earth's orbit around the Sun (100 000 years). These are the famous Milankovic cycles regulating the alternating pattern of Quaternary glacials and interglacials (Fig. 6.17).
2. The long-term (secular) cooling of the deep ocean waters. The $\delta^{18}O$ values for deep-water (benthic) foraminifera reveal a fall of some 10–12 °C since Cretaceous times (Fig. 6.19).

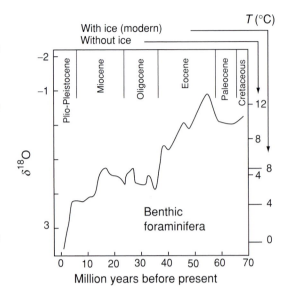

Figure 6.19: The cooling of deep water since the Cretaceous is recorded by the $\delta^{18}O$ value (here reported relative to the PDB standard) of benthic foraminifera. An uncertainty of several degrees results from our ignorance of the volume of polar ice in the Tertiary (3% of the total hydrosphere today): the two temperature scales on the right show two options, one with a volume of ice equivalent to the modern volume, the other without ice (after Miller *et al.*, 1987).

Carbon dioxide is injected into the atmosphere by volcanoes. Erosion then transfers it to the hydrosphere by reactions similar to those described by (6.23) above. It is then removed from the ocean by precipitation of carbonate and by sedimentation in the form of reduced carbon (kerogen, petroleum, and coal). Some of the reduced organic carbon in the sediment is re-oxidized as CO_2 by oxygen in the air when rocks are exposed by erosion; the naturally occurring oil slicks of hydrocarbon rich areas of California and the Arabian Gulf are evidence of this. Human activity is accelerating this process through the burning of fossil fuels in power plants and automobiles. Some of the sedimentary carbonates are removed from the surface system via subduction zones, but we do not know exactly how much of this CO_2 is returned to us by decarbonation at depth and orogenic volcanism in those parts of the world. It appears, however, that the quantity of carbonates at the Earth's surface has increased significantly over geological time.

Climate control by atmospheric CO_2 extends back to the earliest geological times, but there is still debate about the precise mechanisms involved. One hypothesis is that fluctuations in volcanic emissions of CO_2 force the system. In periods of intense volcanic activity, such as the Cretaceous, when rifts were widening very rapidly and volcanic catastrophes were particularly severe, the greenhouse effect was very intense and we picture the luxuriant forests where dinosaurs roamed. It might be thought that an atmosphere with little CO_2 reflects intense primary production in the ocean. It should be possible to detect such conditions through a marked contrast in $\delta^{13}C$ values between shallow

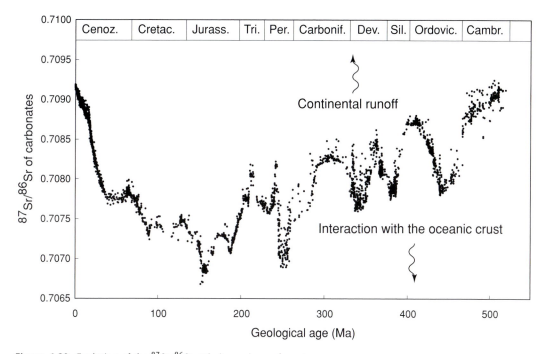

Figure 6.20: Evolution of the $^{87}Sr/^{86}Sr$ ratio in marine carbonates over geological time. The very low Rb content of these rocks ensures that the record of the ratio has not been altered by radioactive ingrowth since deposition. The rapid increase in this ratio since the end of the Cretaceous is attributed to uplift of the high $^{87}Sr/^{86}Sr$ Himalayas, which accelerated the input to the ocean of radiogenic strontium from the ancient basement of the mountain range. This indicates tectonic forcing of the rate of erosion with enormous climatic consequences.

and deep seawater (and the calcite of planktonic and benthic foraminifera living at these levels). However, things are not so straightforward because primary production is apparently restricted by the availability of other elements (phosphorus, iron, and zinc). Abundant CO_2 suggests intense weathering. Intense weathering produces a large influx of alkalinity to the sea and therefore substantial precipitation of limestone. This is a different situation from the earlier one where we were considering CCD levels in glacial times and not actual inputs to the ocean! The Cretaceous was a period of intense limestone sedimentation on continental shelves.

A second hypothesis, suggested by the isotopic geochemistry of strontium in seawater, is that of tectonic forcing. In marine carbonates, the $^{87}Rb/^{86}Sr$ ratio is so low that the $^{87}Sr/^{86}Sr$ ratio has barely varied with time since deposition (see (3.6)). Consequently, $^{87}Sr/^{86}Sr$ remains frozen near its original value and therefore at the $^{87}Sr/^{86}Sr$ value of the seawater (from which the limestone derived) at the time of sedimentation.

The reason for this is that the concentration in the alkaline element Rb (whose behavior is very similar to that of K) is very low compared with that of the alkaline-earth element Sr (whose behavior is very similar to that of Ca). It has been observed that the $^{87}Sr/^{86}Sr$ ratio of seawater as recorded in marine carbonates has increased since the Cretaceous (Fig. 6.20). This increase is evidence of the increased input of continental strontium with its far higher $^{87}Sr/^{86}Sr$ ratio (0.711 on average) than that of strontium derived from the weathering of basalts (0.702). We will see later that this difference reflects the high Rb/Sr ratio of the continental crust. This value points to increased erosion, while the evidence of carbon isotopes and the general cooling of climates and of the ocean since the Cretaceous argues against increased emissions of volcanic CO_2. This suggests, then, more efficient erosional processes, and it has been proposed that the collision between India and Asia, and the resulting uplift of the Himalayas in the Tertiary, were responsible for this. The high elevation of the Himalayas and the monsoon regime that dominates this area favor particularly intense mechanical erosion. Southeastern Asia therefore accounts for a large fraction of the sediment produced over the world. In contrast to the dissolved loads of rivers flowing into the Atlantic, suspended solids and sediments are carried to the Indian Ocean (e.g. along the Ganges and Indus) or to the Pacific (e.g. along the Mekong and Yang-Tse). This second scenario emphasizes the consumption of CO_2 and explains climatic cooling by the more intense erosion resulting from the uplift of the Himalayas.

References

Broecker, W. S. (1994) *Greenhouse Puzzles*. Palisades, Eldigio.

Broecker, W. S. and Peng, T.-H. (1982) *Tracers in the Sea*. Palisades, Eldigio.

Farrell, J. W. and Prell, W. L. (1989) Climatic change and CaCO₃ preservation: an 800 000 year bathymetric reconstruction from the Central Equatorial Pacific Ocean. *Paleoceanogr.*, **4**, pp. 447–466.

Miller, K. G., Fairbanks, R. G. and Mountain, G. S. (1987) Tertiary oxygen isotope synthesis, sea level history, and continental margin erosion. *Paleoceanogr.*, **2**, pp. 1–19.

Petit, J. R., Jouzel, J., Raynaud, D., *et al.* (1999) Climate and atmospheric history of the past 420 000 years from the Vostok ice core. *Nature*, **399**, pp. 429–439.

Sigman, D. M., Altabet, M. A., McCorkle, D. C., Francois, R. and Fisher, G. (1999) The δ¹⁵N of nitrate in the Southern Ocean: consumption of nitrate in surface waters. *Global Geochem. Cycles*, **13**, pp. 1149–1166.

Stallard, R. F. and Edmond, J. M. (1983) Geochemistry of the Amazon. 2. The influence of geology and weathering environment on the dissolved load. *J. Geophys. Res.*, **88**, pp. 9671–9688.

Chapter 7
Mineral reactions

In this chapter we look at the chemical and mineralogical changes accompanying the formation of sedimentary and metamorphic rocks. Marine mud is consolidated into sediment through early diagenetic reactions. These sediments and the other rocks that form the bedrock may be affected by percolating hot water, mineralized to varying degrees, and acids. This is hydrothermal metamorphism. When rock is dragged deep down by subduction and thereby heated and dehydrated, it is transformed, producing a great variety of different mineral assemblages. This process is termed "metamorphism." Some of these thermal processes concern the transformation of organic matter, whose ultimate products are the fossil fuels such as natural gas, petroleum, and coal.

The principal geochemical questions raised by these processes are to identify the nature of the rock before its transformations, the physical conditions (temperature and pressure) of the transformations, the nature and intensity of exchanges between the transformed rocks and the interstitial solutions. Once again the essential analytical tool is thermodynamics and readers may wish to refresh their knowledge of this by referring to Appendix C.

Transformations are often controlled by water pressure and temperature. Let us take the example of the important reaction whereby muscovite (white mica) disappears from gneiss and schist, and which characterizes the entry of metamorphic rocks into granulite facies:

$$KAl_3Si_3O_{10}(OH)_2 + SiO_2 \Leftrightarrow$$
$$\text{(muscovite)} \quad\quad \text{(quartz)}$$

$$KAlSi_3O_8 + Al_2SiO_5 + H_2O \quad (7.1)$$
$$\text{(K-feldspar)} \quad \text{(sillimanite)}$$

The mass action law for near pure solid and gaseous phases (Appendix C) allows us to write:

$$\ln P_{H_2O} = \frac{\Delta H}{RT} + \text{constant} \tag{7.2}$$

where ΔH is the heat (enthalpy) of the reaction. Other reactions involve other gases, such as carbon dioxide, but they are treated in the same way.

The behavior of oxygen is particularly important. The redox reaction between two types of common oxides in igneous and metamorphic rocks:

$$
\begin{array}{ccc}
6Fe_2O_3 & \Leftrightarrow & 4Fe_3O_4 \;\; + O_2 \\
\text{(hematite)} & & \text{(magnetite)}
\end{array} \tag{7.3}
$$

is equivalent to the sum of the two half-reactions:

$$6Fe_2O_3 + 2e^- \Leftrightarrow 4Fe_3O_4 + 2O^- \tag{7.4}$$

$$2O^- \Leftrightarrow O_2 + 2e^- \tag{7.5}$$

This shows that redox reactions are merely a trade in electrons between acceptors (oxidants such as oxygen) and donors (reducing agents such as Fe^{2+}). There is nothing wrong, for example, in considering ferrous iron Fe^{2+} as a complex of ferric ion Fe^{3+} with an electron. As the vast majority of rocks are good electrical insulators, electrons must be conserved throughout mineralogical reactions and phase changes. Appealing to an externally imposed oxygen "fugacity" would be misleading: each and every reduction reaction, i.e. any gain in electrons by one species, must be offset by loss of electrons by another species. No local surplus or deficit of electrons is permitted and species having more than one state of oxidation (above all Fe, C, and S) must engage in balanced electron exchanges. As with water, oxygen pressure can be formulated by the mass action law. For example, for the relationship between hematite and magnetite, we write the equation:

$$\ln \frac{[Fe_3O_4]_{\text{magnetite}}^4 \, P_{O_2}}{[Fe_2O_3]_{\text{hematite}}^6} = \frac{\Delta H}{RT} + \text{constant} \tag{7.6}$$

in which the square brackets indicate the molar fraction of the species in question in each of the solid solutions containing hematite and magnetite, and P_{O_2} is oxygen pressure. Relationships like this can be used either for estimating temperature if the redox state of the system is known, or vice-versa for measuring oxygen pressure if temperature is known. The fugacity of oxygen in many natural rocks is distributed around the famous QFM (quartz–fayalite–magnetite) buffer:

$$
\begin{array}{cccc}
3Fe_2SiO_4 + O_2 & \Leftrightarrow & 3SiO_2 \;\; + & 2Fe_3O_4 \\
\text{(fayalite)} & & \text{(quartz)} & \text{(magnetite)}
\end{array} \tag{7.7}
$$

This equation does not mean that these minerals are present in the rocks, but that the electron balance of the actual mineral assemblage is on average close to that of this buffer.

7.1 Early diagenesis

Mud deposited on the ocean floor normally contains enough dead organic matter to provide food for a burrowing fauna of fish, worms, and mollusks. Even so, almost all these creatures depend on oxygen being available in the interstitial water. Only the top few centimeters of the mud, known as the bioturbated layer, will provide such an oxygenated environment. This layer is recognizable from its burrows and its chaotic character. The top part of the oxygenated sediment is thus mixed to the extent that the finer sedimentary structures disappear. Imagine for a moment that we are observers on the sea floor, resisting burial, and living long enough to observe the sediment sinking beneath our feet (Fig. 7.1). We would see the bioturbated layer acting like a moving average on the sedimentary record (electronics aficionados might liken it to a low-pass filter). If, however, the bottom water is particularly poor in oxygen, usually because of a great abundance of organic matter, burrowing animals cannot breathe, bioturbation is impossible, and the sedimentary record conserves even the finest details down to the infra-millimetric scale (the rocks are then often referred to as laminites).

Below the bioturbated layer, the dissolved oxygen of the interstitial water is fully used up. Other organisms, microbes, take over. To maintain their metabolism they are able to oxidize organic matter from other

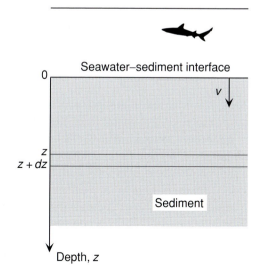

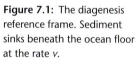

Figure 7.1: The diagenesis reference frame. Sediment sinks beneath the ocean floor at the rate v.

dissolved components, especially by reducing nitrates to nitrogen and sulfates to hydrogen sulfide. These biochemical reactions occur in the space of a few tens of centimeters below the water–sediment interface. If we symbolize dead organic matter (solid) by the very simple compound CH_2O, we can write two relations:

$$CH_2O + SO_4^{2-} \Leftrightarrow H_2S + HCO_3^- \tag{7.8}$$

$$2CH_2O + 2NO_3^- \Leftrightarrow N_2 + 2CO_3^{2-} + 2H_2O \tag{7.9}$$

It can be seen from this how interstitial solutions become increasingly reducing with depth. Under these conditions ferric iron and manganese sedimented on the sea floor as hydroxides are progressively reduced and dissolved in the course of burial (polymetallic nodules are an example of this). Iron precipitates as pyrite FeS_2, a particularly common diagenetic mineral in sedimentary rocks with a high organic matter content, while manganese migrates with pore water and is deposited in other parts of the sedimentary column or is simply expelled again upward into the seawater.

Let us return to our observer on the ocean floor and draw up the balance sheet of matter for one species, say sulfate, between two levels at depth Δz one above the other (Fig. 7.1). The sediment porosity (proportion of interstitial water) is φ and the sulfate concentration in the water is $C(z)$. In the stationary state, i.e. after "some" time, we can write that the difference between the influx of sulfate at z less the outflux at $z + \Delta z$ is equal to the quantity of sulfate destroyed by biological activity. For our observer, the advective flux of sulfate is equal to $\varphi v C(z)$, where v is the sedimentation rate, corrected where necessary for compaction, which progressively expels water from the sediment. Likewise, the diffusive flux of sulfate is $-\varphi D dC(z)/dz$, where D is the diffusion coefficient of sulfate in salt water. We can then write the diagenesis equation, standardizing it to a unit area:

$$\left[\varphi v C - \varphi D \frac{dC}{dz} \right]_z - \left[\varphi v C - \varphi D \frac{dC}{dz} \right]_{z+\Delta z} = -P(z)\Delta z \tag{7.10}$$

where $P(z)$ is the rate of destruction of sulfate by microbial activity. $P(z)$ will be expressed in a suitable form, probably by a first-order kinetic with an appropriate stoichiometric coefficient (see Chapter 4) involving the availability of organic carbon. Solutions to the diagenetic equation are usually combinations of exponentials. By examining (7.10) above, it can be seen that the concentration of sulfate in pore water, like the abundance of organic particles, must decrease with depth: Fig. 7.2 shows the concentrations of sulfate ion in the interstitial water of sediments collected from the Saanich River fjord (west coast of Canada). It could be shown that (7.10) implies exponential decrease in sulfates with the initial sulfate

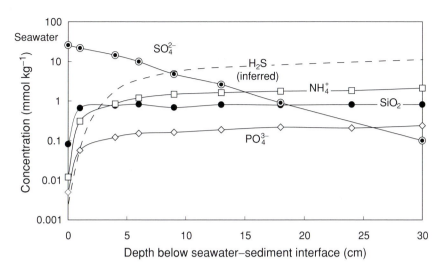

Figure 7.2: Concentration of various species dissolved in interstitial water of the Saanich River fjord (western Canada). With burial, sulfate is reduced to sulfide, organic nitrogen is reduced to ammonia (denitrification), and phosphate and silica of organic origin are remobilized by diagenesis. Data from Murray et al., 1978.

concentration of seawater, which Fig. 7.2 confirms. The reduced sulfur in the form H₂S and the pyrite precipitates increase correlatively, but in this core the sulfur data are missing. Laboratory measurement of diffusion coefficients and estimates of sedimentation rate by radiochronometric methods allow us to determine, by resolving the diagenesis equation, the rate at which sulfate is converted to sulfide and equally the rate of oxidation of the organic matter. Similar behavior is observed for nitrates in seawater. Dissolution of organic matter is reflected by a large increase in ammonia produced by the reduction of nitrogen in proteins and other organic compounds (denitrification). The increase in remobilized phosphate from phospho-organic compounds of soft matter and hard parts of certain organisms (fish teeth, ichthyoliths) is responsible for diagenetic reprecipitation of calcium phosphate (apatite), which may be intense enough to produce phosphate deposits of economic value. The silica remobilized by dissolution of diatom and radiolarian tests is reprecipitated as flint in the enclosing rocks which may be almost fully silicified as cherts. As seawater is very much undersaturated in SiO₂, the expulsion of interstitial water also provides an important source of silica for the ocean.

Until complete, these bacteria-controlled reactions are accompanied by substantial isotopic fractionation, of the order of 25 per mil for carbon and 20 per mil for sulfur, invariably with organic matter having a preference for the lighter isotope.

7.2 Hydrothermal reactions

This term is reserved for all medium temperature reactions, typically from 100 to 500 °C, between aqueous solutions and the rock through which they circulate. These are largely, but not exclusively, hydration reactions. Hydrothermal solutions and ores are the sources of many metals of economic significance. The familiar thermal springs at the surface of the continents give an imperfect sampling of hydrothermal solutions: because the boiling point of hydrous solutions increases from $100\,°C$ at ambient pressure to more than $350\,°C$ at a few kilometers below the surface, the compositions and temperatures of thermal springs do not faithfully represent the properties of the solutions in contact with deep rock. They are the product of intense boiling, to which they are subjected as they rise to the surface. In contrast, the waters of the famous black smokers of mid-ocean ridges, submarine springs whose emergence temperature can reach $400\,°C$ because of the high pressure reigning at the ocean floor (200–450 atmospheres), are better samples of hydrothermal solutions. Many geologically (and economically) important hydrothermal reactions take place at even higher temperatures (400–600 $°C$). High-temperature solutions are often trapped in some minerals, such as quartz, in the form of fluid inclusions.

Balancing hydrothermal reactions requires a combination of cation exchanges between the rock and the solution and the principle of electrical neutrality. First we need to draw up an inventory of relations such as (6.15), which for pure phases, are written:

$$\ln\left(\frac{[Na^+]}{[H^+]}\right)_{solution} = \frac{\Delta H}{RT} + \text{constant} \tag{7.11}$$

At a given temperature, this ratio is constant. The enthalpy ΔH of the reaction varies from one reaction to another. An acid solution (rich in H^+) therefore displaces cations more readily than a basic solution does. This phenomenon is heavily dependent on temperature. Moreover, the sum of cations must remain equal to the sum of anions, which are very often dominated by one or the other, generally chlorine Cl^-, an inert anion i.e. scarcely involved in reactions and whose concentration remains virtually constant. The chemistry of the solution and therefore its capacity to modify the chemistry of the rock depends essentially on all the thermodynamic properties of exchange reactions and on temperature.

For thermometry, it is commonplace to utilize analysis of the water–rock equilibria for hydrothermal reaction thermometry, particularly K/Na fractionation induced by the exchange reaction between feldspar and the solution:

$$K^+ \quad + \quad Na^+ \quad \Leftrightarrow \quad K^+ \quad + \quad Na^+ \tag{7.12}$$
$$\text{(solution)} \quad \text{(feldspar)} \quad \text{(feldspar)} \quad \text{(solution)}$$

Assuming that the feldspar in the reservoir zone of the solution is of the same average composition as the feldspars of granites and metamorphic rocks, the mass action law yields the equation:

$$\log_{10}\left(\frac{[\text{Na}^+]}{[\text{K}^+]}\right)_{\text{solution}} = \frac{908}{T} - 0.70 \qquad (7.13)$$

The sodium and potassium contents of a hydrothermal solution therefore allow an equilibrium temperature with feldspar to be deduced. Solutions can also dissolve minerals without any reaction product: this is the case of silica, the content of which in spa water is used as a thermometer by writing the thermodynamic law of saturation (Appendix C):

$$\frac{\text{d}\ln[\text{SiO}_2]_{\text{solution}}}{\text{d}(1/T)} = \frac{\Delta H}{R} \qquad (7.14)$$

where ΔH is the heat of dissolution of silica in water. By means of a few approximations and by introducing experimental values, the silica thermometer equation is:

$$\log_{10}[\text{SiO}_2]_{\text{solution}} = -\frac{1306}{T} + 0.38 \qquad (7.15)$$

(be wary of the logarithms with different bases). Assuming that there is surplus silica, which is true of most continental rocks, the measurement of the silica content of thermal water gives, with this equation, the equilibration temperature of the solutions with the deep rocks (Fig. 7.3).

From these solubility relationships, we gain some understanding of the mineral associations present in the fractures observed in granitic and

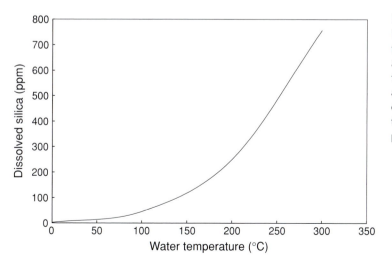

Figure 7.3: Solubility of silica in water. Silica contents of thermal waters can be used as a thermometer to a temperature of about 220 °C, and thus indicate the depth of equilibration. Above this temperature, amorphous silica precipitates as the fluid rises.

metamorphic basements. Pegmatites, characterized by large crystals of albite and K-feldspars, precipitate from relatively high-temperature hydrous fluids (400–600 °C), while the quartz veins, which are commonly associated with ore deposits, form at lower temperatures (200–400 °C).

Water–rock interactions may also leave a profound mark on the isotopic compositions of the geothermal systems associated with magmatic provinces. Rain water being greatly depleted in heavy isotopes of oxygen and hydrogen compared with rock, the $\delta^{18}O$ and δD of magmatic systems (especially granite), which are at the origin of geothermal systems such as that of Larderello in Italy or of the Geysers of California, are generally much lower than those of the starting rocks. This reflects circulation of rain water in rock pores under the influence of heat (convection in a porous medium) and an isotopic exchange between this water and the mineral matrix (Fig. 7.4).

The markedly marine signature of isotopic compositions of oxygen and hydrogen in solutions from black smokers indicates that these formed by seawater infiltrating the edges of the ridges and by reaction at depth with the still hot basalt. Hydrothermal reactions at the mid-ocean ridges play an important role in the magnesium cycle and in controlling the alkalinity of the ocean. It has been observed that water from the black

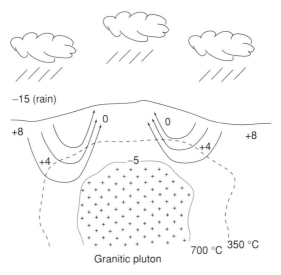

Figure 7.4: Geothermal system controlled by the emplacement of a granite pluton: rain water infiltrates and when heated in proximity to the intrusion becomes lighter and percolates back to the surface. Values shown (in per mil) are those of $\delta^{18}O$ of rock (and of precipitation). Rain water is strongly depleted in ^{18}O and changes the values of rocks, initially here at +8 per mil, to lower values of $\delta^{18}O$, depending on temperature and the water/rock ratio. Crystallization of the granite magmas may release magmatic waters.

smokers of the ocean ridges is particularly acidic, i.e. its pH is much lower (typically 3) than that of deep ocean water (7.6), and that it is totally devoid of magnesium. This can be explained by reactions of seawater with common basalt minerals:

$$\underset{\text{(olivine)}}{Mg_2SiO_4} + \underset{\text{(solution)}}{Si(OH)_4} + H_2O + \underset{\text{(solution)}}{Mg^{2+}} \Leftrightarrow$$

$$\underset{\text{(serpentine)}}{Mg_3Si_2O_5(OH)_4} + \underset{\text{(solution)}}{2H^+} \qquad (7.16)$$

$$\underset{\text{(pyroxene)}}{Mg_2Si_2O_6} + 3H_2O + \underset{\text{(solution)}}{Mg^{2+}} \Leftrightarrow$$

$$\underset{\text{(serpentine)}}{Mg_3Si_2O_5(OH)_4} + \underset{\text{(solution)}}{2H^+} \qquad (7.17)$$

whose effect is to exchange the abundant Mg^{2+} in seawater for protons and therefore to reduce its alkalinity. In such reactions, the silica in the solution comes from dissolution of various enclosing igneous rocks (basalt, gabbro).

The redox reactions between seawater and hot basaltic rock are particularly significant as the state of oxidation of the oceanic crust at the time it is subducted dictates the long-term evolution of the redox state of the mantle. Although the details of redox reactions beneath mid-ocean ridges are complex, they can be represented schematically by oxidation in an acidic medium of the ferrous iron of magmatic rocks offset by reduction of the sulfate in seawater into sulfide ions. The overall reaction breaks down into two half-reactions of oxidation–reduction:

$$Fe^{2+} \Leftrightarrow Fe^{3+} + e^- \qquad (7.18)$$

$$S^{2-} + 4H_2O \Leftrightarrow SO_4^{2-} + 8H^+ + 8e^- \qquad (7.19)$$

By multiplying the first equation by eight and subtracting the second, we obtain:

$$\underset{\text{(seawater)}}{SO_4^{2-}} + \underset{\text{(seawater)}}{8H^+} + \underset{\text{(basalt)}}{8Fe^{2+}} \Leftrightarrow$$

$$\underset{\text{(solution)}}{S^{2-}} + 4H_2O + \underset{\text{(solution)}}{8Fe^{3+}} \qquad (7.20)$$

Ferric ion precipitates as hydroxide $Fe(OH)_3$ and the sulfide as sulfide minerals of iron and copper (chalcopyrite), and zinc (sphalerite) forming the mineralized submarine chimneys of the black smokers. Sulfur isotopes indicate, however, that, within the mid-ocean ridge hydrothermal systems, sulfur derived from seawater through the previous reaction is dominated by sulfur leached from the basaltic rocks. The sulfate-reduction reaction also transforms part of the ferrous iron of the oceanic crust into ferric iron stored as oxides or silicates (epidote). Subduction of the oceanic crust therefore controls not only mantle hydration but also its relative proportions of ferric and ferrous iron, and its electrical conductivity. The Earth's mantle, into which enormous amounts of oxidized crust have been recycled over the Earth's history, is thus far more oxidized than the Moon's mantle.

7.3 Metamorphism

Metamorphic transformations affect all rocks drawn down deep into the Earth by subduction. They are particularly perceptible where continents collide, i.e. when mountain ranges form. In contrast to hydrothermal reactions, these are largely, although not exclusively, dehydration reactions under the effect of temperature or excess CO_2. Metamorphic facies correspond to given temperature and pressure ranges and these are usually bounded by specific mineralogical reactions: the greenschist facies (250–450 °C), the amphibolite facies (450–700 °C), and the granulite facies (>700 °C) correspond to increasing temperatures at usual pressures; at higher pressures (>30 km) we speak of blueschist facies, and at higher temperatures of eclogite facies.

The nomenclature of metamorphic rocks, based on their mineralogy, is fairly straightforward. Gneiss contains feldspar and quartz with variable proportions of other minerals and is frequently similar to granite in chemical composition. Schist contains little or no feldspar and is typically composed of quartz and mica; it is of similar composition to claystones. Amphibolite contains amphibole, with or without plagioclase feldspar; it is similar in composition to basalt. Most dehydrating metamorphic reactions could be described by dehydration reactions of the type described by (7.2) and represented by straight lines in a plot of $\ln P_{H_2O}$ vs $1/T$ K, but it has become customary to represent metamorphic equilibria as curves on a simple pressure–temperature graph (T °C, P_{H_2O}). A metamorphic grid of this sort is commonly used to determine the temperature and pressure conditions prevailing in ancient metamorphic environments. Oxidation–reduction reactions are also used to determine temperature and oxygen pressure. Other reactions, finally, do not involve any fluid, as for example the polymorphic transformation of aluminum silicates:

$$Al_2SiO_5 \Leftrightarrow Al_2SiO_5 \Leftrightarrow Al_2SiO_5 \qquad (7.21)$$
$$\text{(andalusite)} \quad \text{(kyanite)} \quad \text{(sillimanite)}$$

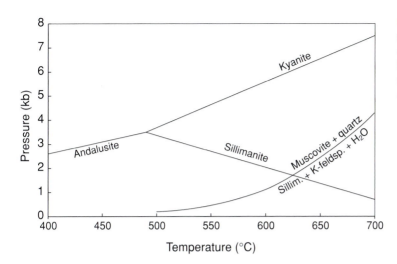

Figure 7.5: Metamorphic grid showing dehydration reaction curves that can be described by (7.2) and three solid–solid reactions for which the slopes are virtually constant.

Such reactions, not usually directly involving fluids, are represented by straight lines in pressure–temperature diagrams (Fig. 7.5). The position of the Clapeyron curve of these different reactions in pressure–temperature space must be carefully calibrated by experiment or obtained by thermodynamics (see Appendix C).

At this stage we should look more closely at the relationships between fluid pressure – the fluid for convenience we take to be water – and the pressure of the surrounding rock. As water is three times less dense than rock, the weight of a column of water is three times less than the weight of a column of rock of the same height. Near the surface, the pores are interconnected and interstitial fluid pressure is therefore equal to hydrostatic pressure. The weight of the water column being about one-third the weight of the rock column, pressure in the rock matrix (lithostatic pressure) is higher than that of interstitial water. With depth, the rock is compacted and the pores tend to close, progressively isolating the pore water a few kilometers beneath the surface. Fluid pressure and rock pressure are then in equilibrium. When the temperature of this unit is raised by a metamorphic event, as the water has greater thermal expansivity than rock, it is at higher pressure than the rock matrix. As the tensile strength of the rock is generally low, this pressure contrast causes hydro-fracturing and water escapes.

Metamorphic dehydration reactions entail a loss of water but also of dissolved solids, and metamorphic rocks must be considered as open systems where very large quantities of water may have circulated. During this process, the chemical composition of the initial rock is transformed. As metamorphic reactions proceed, the distribution of trace elements and chronometric systems are severely disrupted but, because the system is open to fluid circulation, this rarely happens in a predictable way. Since, at the temperatures at which metamorphism occurs, ^{18}O tends to become enriched in the fluid (contrary to what happens at ordinary temperatures), the isotopic composition of the oxygen of rocks drifts toward lower $\delta^{18}O$ levels closer to those of the mantle. At temperatures in excess of $500\,°C$, the oxygen isotope fractionation coefficients tend to unity, there is little water left to exchange oxygen with, and isotopic changes become less important.

Extreme metamorphic conditions cause rocks to melt when they have a high water content; the hydrated melting of common crustal metamorphic rocks is referred to as anatectic melting. When water is absent or the dominant fluid is CO_2, granulite facies conditions pertain. Circulation of CO_2 promotes migration and the loss of elements that are normally inert under hydrated conditions: it is known that granulite facies rocks, which are so common at the base of the continental crust, have lost much of their uranium and some of their thorium. The production of heat in granulites of the deep part of the continental crust by the radioactive elements U, Th, and K is therefore normally very low, with, as a consequence, a distinctive isotopic composition of their lead, which is normally unradiogenic.

7.4 Water/rock ratios

The motivations for finding how much water a given rock sample has "seen" during diagenetic, hydrothermal, or metamorphic processes are varied: we may want to assess the reserves of a particular element of economic importance remaining in the source rock of a particular ore deposit, we may need to know how far water circulation affected the thermal regime of a given area, or we may worry that too much circulating water disturbed the initial geochemical properties of a rock, a mineral, or fossil remains. In order to provide a quick estimate of the extent of water–rock interaction, geochemists often refer to the concept of a water/rock ratio. Given a set of geochemical observations on an altered rock or a diagenetic/hydrothermal solution, geology may often give some hints at what the untransformed rock may have looked like (a basalt, a granite), while other constraints may be good enough to let us infer the geochemistry of the reacting fluid (meteoritic water at a given latitude, for instance). A number of geochemical properties can be used, concentrations and isotopes, but probably the most popular are $\delta^{18}O$ and $^{87}Sr/^{86}Sr$. Let us suppose, for instance, that we measured the $\delta^{18}O_{HR}$ of a hydrothermally altered basalt sample to be -2 per mil, while examination of thin sections suggests that the rock was hydrothermally altered at $350\,°C$ in the greenschist facies. Paleogeography also suggests that when this particular basalt was erupted, the latitude was such that meteoritic water had a $\delta^{18}O_{MW}$ of -10 per mil. We will assume that the $\delta^{18}O_{FR}$ of the fresh basalt was in the range of mantle-derived magmas, say $+5.5$ per mil. Can we find out how much water interacted with this particular basaltic sample before it turned into a metabasalt?

We will note R and W the mass of rock and water reacting with each other and assume that the amount of material exchanged during the reaction is small (R and W are constant). We will also neglect the fact that water and rock contain slightly different oxygen concentrations. We can write the isotopic mass balance during the reaction as

$$R\delta^{18}O_{FR} + W\delta^{18}O_{MW} = R\delta^{18}O_{HR} + W\delta^{18}O_{HW} \tag{7.22}$$

where HW refers to the hydrothermal solution, which, unfortunately, can no longer be sampled and analyzed. Rearranging, we get the water/rock (W/R) ratio as:

$$\frac{W}{R} = \frac{\delta^{18}O_{FR} - \delta^{18}O_{HR}}{\delta^{18}O_{HW} - \delta^{18}O_{MW}} \tag{7.23}$$

It is often observed that, for mass balance reasons, the $\delta^{18}O$ value of igneous whole-rocks is similar to that of their feldspar. This property allows us to estimate the $\delta^{18}O$ of the hydrothermal water from the $\delta^{18}O$ of the altered rock and the fraction factor of $^{18}O/^{16}O$ between feldspar and water at 350 °C, i.e:

$$\delta^{18}O_{feldspar} = \delta^{18}O_{HW} + 4 \tag{7.24}$$

Rearranging, we get the water/rock (W/R) ratio as:

$$\frac{W}{R} = \frac{5.5 - (-2)}{(-2) - 4 - (-10)} = 1.875 \tag{7.25}$$

Water/rock ratios in the range 1–5 are very common, while much higher values (up to several hundred) are not exceptional, especially during diagenesis. Examples based on Sr isotopes may be constructed in a similar way.

Reference

Murray, J. W., Grundmanis, V. and Smethie, W. M. (1978) Interstitial water chemistry in the sediments of Sannich Inlet. *Geochim. Cosmochim. Acta*, **42**, pp. 1011–1026.

Chapter 8
The solid Earth

Before discussing the formation of the major geological units of the solid Earth, we should review the internal structure of our planet as described by seismic wave studies (Fig. 8.1). The most important discontinuities observed by seismologists down to the base of the mantle are:

- The base of the crust (called the Mohorovičić discontinuity or Moho), 40 km below the continents, but only 5–7 km beneath the oceans.
- The base of the lithosphere, on average 80 km below the oceans, and deeper still beneath the continents. This is the lower boundary of the rigid tectonic plates. The softer part of the upper mantle underneath the lithosphere is called the asthenosphere.
- The 440 km discontinuity corresponding to a change in olivine structure (spinel or ringwoodite phase).
- The 660 km discontinuity corresponding to the transformation of all minerals into perovskite and minor Fe-Mg oxide (magnesio-wüstite). This is the base of the upper mantle.
- The mantle–core boundary at about 2900 km. Above this boundary is a seismically abnormal layer some 200 km thick, known as the D'' layer.

The core is divided into a liquid outer core, responsible for the Earth's magnetic field, and a solid inner core. The core is metallic and composed mostly of iron and nickel.

Plate tectonics is a powerful theory that unifies the geological expression of crustal and upper mantle geodynamics (Fig. 8.2). The Earth's surface is covered with rigid lithospheric plates that may or may not carry continents. These plates are about 80 km thick beneath the oceans and at least twice as thick beneath the continents. They move apart along

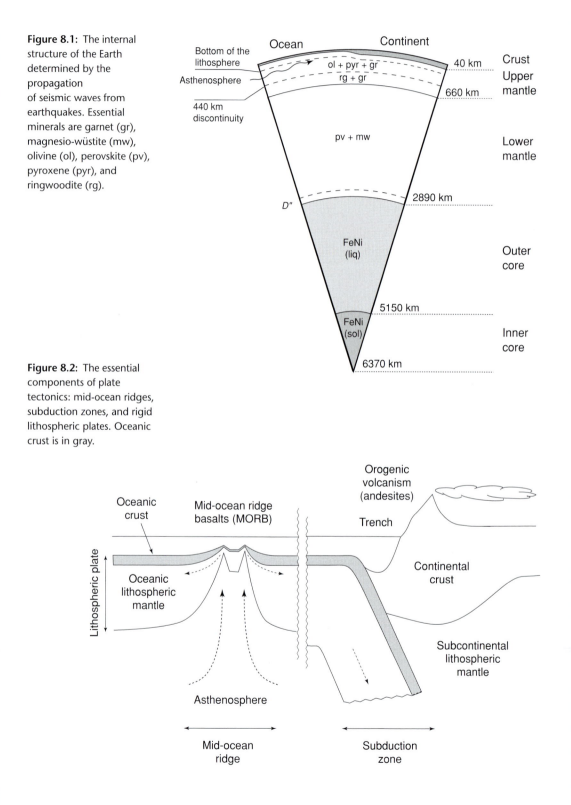

Figure 8.1: The internal structure of the Earth determined by the propagation of seismic waves from earthquakes. Essential minerals are garnet (gr), magnesio-wüstite (mw), olivine (ol), perovskite (pv), pyroxene (pyr), and ringwoodite (rg).

Figure 8.2: The essential components of plate tectonics: mid-ocean ridges, subduction zones, and rigid lithospheric plates. Oceanic crust is in gray.

mid-ocean ridges and converge along subduction zones (marked by the deepest ocean floors, the trenches), where one slides under the other.

Plate tectonics is not a separate mechanism from mantle convection, but is merely its surface expression. The mantle is not a molten medium: like ice flowing in glaciers, the mantle is a solid that is nonetheless deformable when worked rather slowly. Convection is a generalized movement of the mantle maintained by density inversions (heavy above light) brought about by thermal contrasts within the Earth: a material is lighter when hot than when cold. But also hot material deforms more easily than cold material, so that radioactive heating not only makes the mantle hotter and lighter but softer, which facilitates convective movements. The temperature dependence of viscosity therefore acts as a thermal regulator for the thermal regime of the mantle. Hot mantle material pours out from under the mid-ocean ridges, while cold lithospheric plates are injected deep into the mantle by subduction. Mantle convection therefore has the effect of extracting heat by replacing deep hot mantle by material that was cooled next to the surface. Since thermal conduction through the lithospheric plates is a rather inefficient cooling mechanism, it is not exaggerated to say that the mantle cools from below thanks to the subduction of young cold plates.

Density inversions are maintained by heat from radioactive elements (U, Th, K) contained in the mantle but also by the heat released from the core. In this case, fluid mechanics teaches us that convection in a medium heated from the bottom is unstable, with irregular spurts of hot, less dense material rising rapidly to the surface: these are hot spots or volcanic plumes that begin catastrophically at the core–mantle boundary. The heads of these instabilities produce gigantic eruptions covering areas larger than Texas with lava in comparatively short geological times (about a million years) and may break up whole continents. The separation of North America and Europe or of Antarctica and Australia is attributed to this process. The residual plume tail may remain active for tens of millions of years and, as plates move across the surface, it may form strings of volcanoes thousands of kilometers long such as those of Hawaii. Where these plates are not continental, cooling makes them progressively denser and they sink into the mantle as planar structures at subduction zones. The cold parts constantly drag the remainder of the plate down into the mantle where it tears apart at its thinnest points. This sinking is offset by the formation of young lithosphere along the scars, the mid-ocean ridges, rather like a tablecloth slipping over the edge of the table from the center of which it is being constantly woven and fed out. The subduction zones are the downgoing limbs of mantle convection, while material rises from the deep either with the general circulation (ca. 90%) or as plumes (ca. 10%).

The continents are formed by relatively light felsic material and behave like corks fixed within the lithospheric plates. Mountain ranges form when two continental "corks" collide: a fraction of the continental rocks may then be carried down to very great depths (up to 200 km). There, they are subjected to very high temperatures and pressures and their mineral assemblage becomes eclogitic. This is the origin of high-pressure metamorphism.

The most significant chemical fractionation processes within the Earth are the formation of oceanic lithosphere, hot spots, and continental crust. The fractionation processes attending melting and crystallization – the essential mechanisms by which magmatic rocks originate – are responsible for the diversity of chemical composition of such rocks. They also determine indirectly the geochemical variability of metamorphic rocks and clastic sedimentary rocks (clays, sands). The essential mixing processes are associated with hybridization (or contamination) of magmas and mantle convection. The current state of the mantle and continental crust is the outcome of competition among all of these processes.

8.1 The geochemical variability of magmas

Magma, the common term for a molten rock, may contain crystals in suspension, usually called phenocrysts. Molten magma reaching the surface is known as lava. If the magma crystallizes completely, most commonly as an effect of slow cooling, the rock produced is said to be intrusive or plutonic. If cooling is too fast for crystallization to be completed, for example in a submarine eruption, the liquid is quenched to a glass. After cooling, it forms an effusive or volcanic rock.

The two most abundant types of magmatic rocks are basalt in ocean areas and granite in continental areas. A basalt reaches the surface at a temperature of 1150–1250 °C, and a granitic liquid (rhyolite) at about 1000 °C. Table 8.1 gives the characteristic major element compositions of magmatic rocks in their standard form, i.e. by oxide weight. Basalt is rich in iron, magnesium, and calcium, whereas granite is rich in silica and alkaline elements. We will now look at the two mechanisms responsible for the variability of magmatic rocks, melting of the mantle and crust, and magmatic differentiation.

8.1.1 Melting of the mantle and crust

The principal cause of chemical variability of the primary melts extracted from the mantle and the crust is the nature of the source rock. By primary, it is meant that the magma did not crystallize mineral phases, a process

Table 8.1: *Chemical composition of a few representative rocks, basalts (tholeiitic or alkali basalt), andesite, and granites (in weight percent oxide)*

Locality	Rock	SiO$_2$	TiO$_2$	Al$_2$O$_3$	FeO	MgO	CaO	K$_2$O	Na$_2$O	P$_2$O$_5$
Mid-ocean ridge	Thol. bas.	50.9	1.2	15.2	10.3	7.7	11.8	0.1	2.3	0.1
Hawaii	Alk. bas.	45.9	4.0	14.6	13.3	6.3	10.3	0.8	3.0	0.4
Hawaii	Thol. bas.	49.3	2.8	13.4	11.5	7.7	11.0	0.4	2.3	0.3
Siberia (trap)	Thol. bas.	48.7	1.4	14.8	10.3	7.3	9.7	0.9	2.4	0.1
Cascades	Andesite	60.4	0.9	17.5	6.4	2.8	6.2	1.2	4.3	0.2
Australia	Granite (I)[a]	69.5	0.4	14.2	3.2	1.4	3.1	3.2	3.5	0.1
Australia	Granite (S)[b]	70.9	0.4	14.0	3.2	1.2	1.9	2.5	4.1	0.2

[a,b] See p. 143 for a definition of I- and S-type granites.

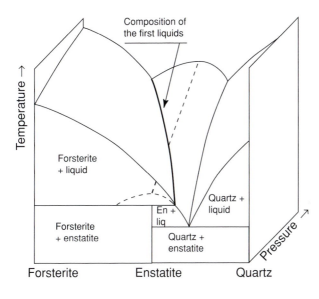

Figure 8.3: The effect of pressure on the composition of the first melt (bold line). With depth, peritectic reactional melting, producing basalts with a fairly high silica content (tholeiites), gives way to eutectic melting forming more olivine-rich basalts (after McBirney, 1992).

called fractional crystallization and which will be discussed in the next section. Melting of the peridotitic upper mantle, composed mainly of olivine, pyroxene, and aluminous minerals (feldspar, spinel, and garnet) yields basalts. Melting of the continental crust, formed by wet sediments accumulated by erosion, and by their metamorphic equivalents (schist and gneiss) produces granitic liquids. A second factor of variability is pressure, i.e. the depth of melting (Fig. 8.3). At shallow depths, melting of the mantle produces tholeiitic basalts, i.e. basalts that are somewhat richer in Si and poorer in Mg than other types of basalts. At greater depth the trend is reversed and magma is richer in Mg and Fe and poorer in Si: these are alkali basalts. The third factor of variability is temperature or, equivalently for a system of a given composition at a given pressure, the degree of melting: the major elements that enter the melt most readily (Si, Al, Ca, K, Na, Ti) and incompatible trace elements (Rb, Zr, Ba, rare-earths, Th, U, etc.) are relatively abundant in the first melts. As melting proceeds, the liquid becomes richer in refractory elements (Mg, Cr), fresh melt dilutes incompatible elements, and basalt gives way to picrite. Melt temperature and composition are not independent. Close to the minimum eutectic temperature, the melt fraction is small and the liquid composition is buffered by the residual minerals. This is the case not only of mantle melting at high pressure (Fig. 8.3), but also of the crust. At higher temperatures, melting progressively consumes residual minerals, and the liquid composition drifts toward that of the initial solid. In the mantle, an increase in temperature and degree of melting produces Mg-rich rocks. Komatiites are high magnesian lavas found only in Archean terrain (except for the small Tertiary flows on Gorgona Island). They are explained by very intense melting of the mantle of the order of 50% and are thus evidence of the very high mantle temperature during the earliest ages of the Earth.

The fourth factor governing the composition of melts is the water and carbon dioxide content of the source rock. The effect of these volatile compounds is similar to that of elements with low melting points (e.g. alkali elements). The presence of such "impurities" lowers the melting point of the rock (Appendix C, (C.32)). As long as a gas phase is absent, water behavior is similar to that of an ordinary incompatible element, and this is also true for any other gaseous species (Co_2, Ar, He). It is noteworthy that the H_2O/Ce ratio is practically constant at 300 in basaltic magmas from different environments (mid-ocean ridges, oceanic islands, subduction zones) when measurements are made on melt inclusions within crystals, i.e. from non-degassed samples. As water is nonetheless the most abundant impurity, it soon reaches saturation and forms a vapor phase, as the magma rises, and separates out from the melt. From this point, if the vapor phase is mostly water, the mechanical equilibrium of vapor and the surrounding rock requires the water pressure to be equal to the pressure of the surrounding rock and the water concentration of the magma is primarily controlled by pressure. The presence of abundant water also changes the composition of the melt: a water-rich source rock produces a magma that is richer in silica than a dry rock. Finally, the abundance of minerals with a low melting temperature affects the capacity of the rock to produce magma (melt productivity). At a given temperature, a mantle rock rich in pyroxene and aluminous minerals will produce more liquid than a peridotite with a very high refractory olivine content.

Geochemical observations of lavas, notably of trace elements, can be interpreted quantitatively with equilibrium melting or fractional melting models (Eqs. (2.17) and (2.31)). More sophisticated models, mentioned earlier, draw on the percolation of fluids through rock pores or the compaction of molten rock. These provide a more realistic picture of the geochemistry of melt products. The mantle melting mechanisms themselves are still poorly understood. The liquid first appears at the grain boundaries. Because it is less dense than the solid mantle, buoyancy tends to expel the melt upwards as would a sponge full of water under its own weight. So long as the porous matrix can be deformed, liquids percolate, producing chromatographic fractionation effects. When they penetrate the colder, more rigid parts of the mantle, they collect in fractures and are channelled quickly toward the surface forming volcanic systems.

Melting in the continental crust, most often in the presence of water, produces granitic liquid: the composition of the melt is buffered by a mineralogical assemblage of quartz and feldspar (eutectic melting), so that the variability of granite composition is fairly narrow. The presence in the granite source of sedimentary material or material weathered at low temperature is attested to by many indicators, the most obvious being the commonly high level of their $\delta^{18}O$ value (8–14 per mil). During orogenic periods, substantial crustal melting may occur: if we imagine that a granite intrusion, some 150 km long, 50 km wide and 6 km thick (a common enough dimension in many shield areas), represents a partial melt rate of 30%, there must have been an enormous molten layer of nearly 20 km of crust at the time these granites formed. Crustal melting is therefore a large-scale process. Narrow swarms of granitic intrusions, occasionally elongated over thousands of kilometers (batholiths), often mark the edge of subduction/collision zones: hybrid magmas formed by mixing of mantle-derived basalts and andesites with felsic anatectic melts dominate these structures.

Some granites, known as S-type granites, form primarily when sediments and their metamorphic equivalents melt deep in the continental crust. Their high $\delta^{18}O$ values (9–15 per mil) imply that their source rock formed at low temperature in the presence of water. The abundance of muscovite and the low Fe^{3+}/Fe^{2+} ratio of the S-granites indicate that this source contained large proportions of oxygen-starved clay minerals. Their high $^{87}Sr/^{86}Sr$ and low $^{143}Nd/^{144}Nd$ ratios reflect the fact that this source has evolved over hundreds of millions, even billions of years, with a high Rb/Sr ratio typical of metasedimentary rocks. In contrast, other granites, known as I-type granites, have low $\delta^{18}O$ values (6–9 per mil) indicative of a source dominated by igneous rocks that were only slightly affected by low-temperature alteration (typically older plutonic rocks). Ubiquitous hornblende reflects the abundance of Na and Ca in the source rock. More than 25% of the iron is oxidized, while Sr is less radiogenic and Nd more radiogenic than in S-granites: all these features point to the melting of deeper levels of the continental crust.

Granitic magmatism is almost exclusively limited to the Earth's continents: a sample of lunar granite has been described, a few lavas of granitic composition are known in Iceland and the islands of Réunion and Ascension, and on mid-ocean ridges, but their isotopic characteristics (O, Sr, Nd) make them more like basalts than continental granites. These granitic rocks must have formed by remelting of the basaltic pile, although, in rare cases, extreme differentiation of basaltic magmas may also leave felsic lavas as the end-product.

8.1.2 Differentiation of magmatic series

A separate cause of magma chemical variability is differentiation, already referred to at the end of Chapter 2. As magmatic liquids are less dense, at least at upper mantle pressure, they are subjected to the buoyancy forces of the surrounding rocks and therefore rise, initially by percolation when the matrix is deformable and then, in the lithosphere, along fractures. As they rise, they cool by contact with the walls and the solubility limit is reached for each mineral in turn. In basalts at high pressure, olivine saturation is reached first, followed by that of clinopyroxene and, finally, plagioclase. If cooling occurs at low pressure (<15 km), as under the mid-ocean ridges, the order of appearance of plagioclase and pyroxene is reversed. In some cases, an intermediate reservoir may temporarily accommodate the magma. This is the mythical magma chamber, a concept that pervades petrological literature. Such magma chambers are actually uncommon in island volcanoes, but are ubiquitous beneath ridges. This contrasting pressure-dependent behavior is well illustrated by the evolution of the major element concentrations in various types of basalts. Most ocean island basalts (except from those islands that lie on the ridge axis, such as Iceland, or adjacent to it, such as the Galapagos) differentiate at medium pressure and remain undersaturated in plagioclase: when plotted against a differentiation index, such as FeO/MgO or even MgO, the Al_2O_3, or TiO_2 concentrations attest to the essentially incompatible behavior of Al and Ti (Fig. **??**) and increase in differentiated melts. In contrast, mid-ocean ridge basalts fractionate plagioclase: Al_2O_3 is compatible and decreases, while FeO and TiO_2, rejected by plagioclase, remain incompatible and increase in residual melts. In arc (orogenic) magmas, the relatively high water content suppresses early plagioclase saturation: Al, Ti, and Fe are controlled by clinopyroxene and olivine fractionation.

The minerals thus formed may be denser (olivine and pyroxene) or less dense (plagioclase) than the magma from which they crystallize. The

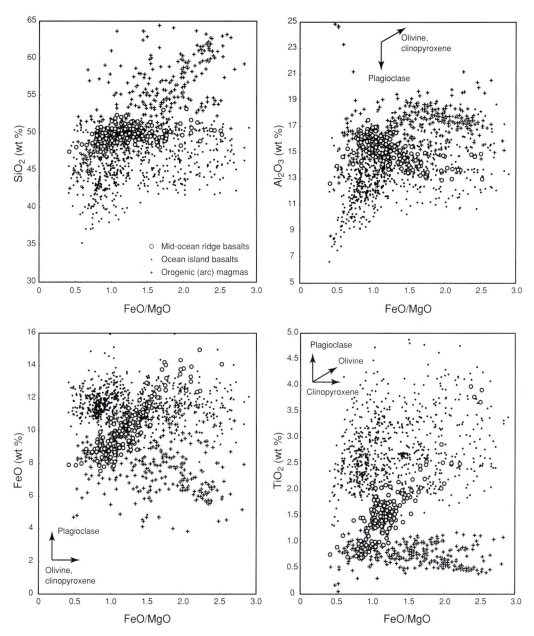

Figure 8.4: Changes in the behavior of some major elements in basalts from different tectonic settings as a function of plagioclase saturation. The arrows represent the effect of removing a particular mineral on the major element content of the residual melt. As shown by the evolution of SiO_2, the x-axis coordinate (FeO/MgO) represents the extent of differentiation: lavas become more differentiated from left to right. Plagioclase saturation in MORB as a result of low-pressure (< 5 kb) differentiation makes Al compatible and Fe and Ti more incompatible. Plagioclase undersaturation as a result of high pressure fractionation (OIB) or high water pressure (arc magmas) has the opposite effect: Al becomes incompatible in contrast with Fe and Ti, which become more compatible.

denser ones tend to sink and sediment out, while the light plagioclase floats, with cumulates forming in both cases. Such crystallization followed by the separation of minerals selectively removes elements from the parent magma: this is magmatic differentiation by fractional crystallization. This process is responsible for the formation of magma suites of very variable composition, for example the basalt–phonolite–trachyte suite, fine examples of which can be seen in composite volcanoes such as in the Canary or the Society Islands. The question of the chemical evolution of these suites is generally well understood. It is represented by situations of fractional crystallization (Rayleigh's law), possibly complicated by the assimilation of surrounding rocks (AFC = assimilation and fractional crystallization), or recurrent episodes of reservoir replenishment. Compatible elements (Ni, Cr, Sc, Sr) will be good tracers of fractional crystallization as they vary greatly for a low degree of fractionation. Nickel indicates the precipitation of olivine, Sc that of clinopyroxene, Eu and Sr that of plagioclase. It can be calculated, for example, from (2.30), that a modest 10% precipitation of olivine, for a partition coefficient of Ni between olivine and liquid of 15, lowers the parent magma content of this element by 77%. An excess Eu with respect to the neighboring rare-earth elements Sm and Gd will signal that a mafic rock is a plagioclase cumulate (gabbro), while a deficit requires plagioclase removal. In contrast, incompatible elements will be uniformly enriched in the residual liquid. Their relative distribution carries little or no information about magmatic differentiation: fractionation of 10% olivine enriches the magma in both Th and La (zero partition coefficient) by 11%. In contrast, the variability in incompatible elements induced by partial melting is very high compared with that induced by fractional crystallization. As already alluded to, incompatible elements are therefore particularly useful for the study of partial melting.

In the more viscous granitic liquids, gravity separation of minerals is less efficient than in the normally much more fluid basalts. Granitic rocks are generally far more difficult to interpret than basalts in terms of geochemistry as they are plutonic rocks and fully crystallized: it is practically impossible to assert from examination of a sample of granite whether it is a sample of congealed liquid magma, the cumulates derived from it, or some intermediate combination of both.

8.2 Magmatism of the different tectonic sites

Figure 8.5 schematically shows the geographical distribution of various magma types in relation to their geodynamic site.

The temperature at which rock melts increases with pressure and decreases sharply in the presence of water. The low water content of

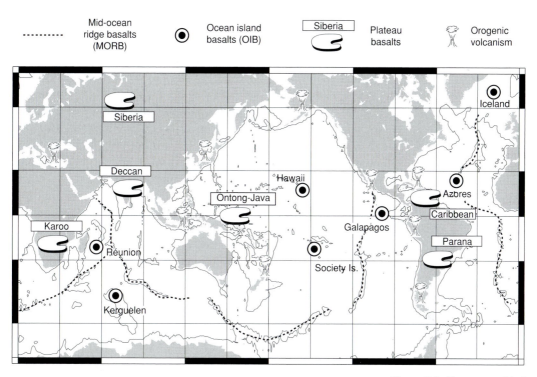

| · · · · · · · · · · | Mid-ocean ridge basalts (MORB) | ◉ | Ocean island basalts (OIB) | Siberia | Plateau basalts | | Orogenic volcanism |

Figure 8.5: The geographical distribution of different types of volcanism. To bring out the shallower parts of the ocean (ridges and plateaus) the -3000 m depth contour is shown (drawn using GMT).

the mantle (about 500 ppm) limits its effect on upper-mantle melting. Mechanical effects of trace amounts of water, however, standout: dry olivine is stiffer than wet olivine and melting leaves behind a fairly rigid residue. The production of lava is associated with decompression of the mantle in plumes and beneath oceanic ridges but also with dehydration of wet material (weathered sediments and basalts) sinking in subduction zones. These three sites produce basalts, i.e. lavas rich in Mg, Ca, and Fe and poor in Si. Other types of magma are associated with continental collision zones: these are felsic magmas rich in Si and poor in Mg, whose characteristic product is common granite. Continental margins are often associated with mixed magmatism, which goes by the name of orogenic magmatism. Examples are the deadly eruptions of Mount Pelée, Mount Pinatubo, or El Chichon. A common type of lava erupted by these volcanoes is andesite.

The nature of the mantle, which produces different types of basalt, is an active area of research. The greatest progress has been accomplished recently for the ocean island basalts (OIB) that form chains of volcanoes associated with hot spots, and the mid-ocean ridge basalts (MORB) that form the ocean floor. MORB production is connected with decompression of the upper layers of mantle, forming the asthenosphere, which rise beneath the mid-ocean ridges.

MORBs are extremely depleted in highly incompatible elements such as Ba, Th, K, and light rare-earths compared with what might be expected of the melting of unprocessed mantle peridotite: the asthenospheric mantle had already lost a large proportion of its most readily fusible elements before the present-day MORB formed. This can be shown easily by comparing two trace elements with different degrees of compatibility, e.g. two elements of the rare-earth family, ytterbium (Yb) and lanthanum (La). If F is the fraction of melt prevailing during formation of MORB and $D_{s/l}^{Yb}$ the partition coefficient of Yb between solid residue and melt, (2.17) can be written:

$$C_{MORB}^{Yb} = \frac{C_{source}^{Yb}}{F + D_{s/l}^{Yb}(1 - F)} \tag{8.1}$$

As a first approximation, La is almost completely incompatible and its solid residue/melt partition coefficient $D_{s/l}^{La}$ is so small that it can be taken as zero. We can then approximate:

$$C_{MORB}^{La} = \frac{C_{source}^{La}}{F} \tag{8.2}$$

Extracting F from (8.2) and replacing it in (8.1) gives:

$$\left(\frac{C^{La}}{C^{Yb}}\right)_{MORB} = \left(\frac{C^{La}}{C^{Yb}}\right)_{source} \left(1 + D_{s/l}^{Yb}\frac{1-F}{F}\right) > \left(\frac{C^{La}}{C^{Yb}}\right)_{source} \tag{8.3}$$

Because the term in the parentheses of the right-hand side is greater than unity, this equation shows that the La/Yb ratio must be greater in the magmatic liquid (MORB) than in the source mantle. This approach is valid, of course, for all trace elements. Figure 8.6 shows the distribution of rare-earth elements in melts from the mantle and in their residues for different degrees F of melting. The enrichment of incompatible elements in the melts with respect to their mantle source is visibly much greater for the highly incompatible elements (La, Ce) than for the less incompatible elements (Yb, Lu). The inescapable conclusion is that MORB depletion in highly incompatible elements, such as La, Ce, but also Th and Ba, compared with more compatible elements such as Sr, Yb, and Ti (see Fig. 2.7), implies that their source is depleted in these highly incompatible elements to an even greater extent.

Radiogenic isotopes show that depletion of the mantle under the mid-ocean ridges in incompatible elements is an ancient phenomenon. The highly incompatible radioactive isotope [87]Rb is greatly diminished compared with the more compatible radiogenic isotope [87]Sr, and in the course of geological time a relatively low [87]Sr/[86]Sr ratio of the asthenospheric mantle results (0.7025). By contrast, the parent isotope [147]Sm is more compatible than its daughter isotope [143]Nd. It is therefore enriched in residual mantle by comparison with the melt. In the course of time, the asthenospheric mantle acquires a relatively high [143]Nd/[144]Nd ratio

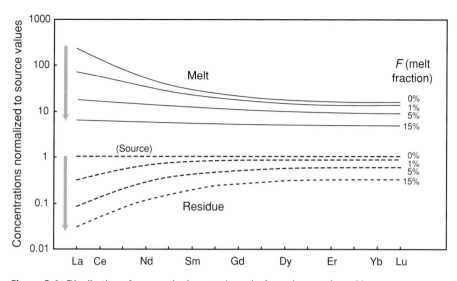

Figure 8.6: Distribution of rare-earth elements in melts from the mantle and in their residues for different degrees F of melting. Concentrations were calculated using equilibrium batch melting equations similar to (8.1) and then normalized to the values in the mantle source. The residues were assumed to contain 15% clinopyroxene with the following mineral–liquid partition coefficients: La: 0.03, Ce: 0.085, Nd: 0.19, Sm: 0.29, Dy–Yb: 0.44 and 85% of rare-earth free olivine and orthopyroxene. Note that the highly incompatible elements (La, Ce) are extracted much more efficiently from the residues than the more compatible elements (Yb, Lu). Since MORBs are depleted in La, Ce with respect to Yb, Lu (see Fig. 2.7), their sources are necessarily even more depleted, which demonstrates that they went through previous melt extraction events. This is why the mantle under the mid-ocean ridges is usually referred to as the depleted mantle (DM).

(see (3.23)) of the order of 0.5131. The ^{187}Os/^{188}Os ratio evolves like ^{143}Nd/^{144}Nd, but in a more extreme way.

The values of the isotopic ratios of present-day magmatic rocks are usually plotted in a series of binary diagrams, the most popular being undoubtedly the plot of ^{143}Nd/^{144}Nd versus ^{87}Sr/^{86}Sr (Fig. 8.7). The geochemical jargon "depleted" simply refers to a deficit of incompatible (or fusible) elements with respect to compatible (or refractory) elements against an ideal average mantle model (Bulk Silicate Earth or BSE). The opposite of "depleted" is of course "enriched." Depletion can be determined for the time immediately before melting using the analysis of trace elements. It may also be determined for long periods (1–3 billion years) that precede melting by analyzing isotopic compositions of elements containing a radiogenic isotope (Sr, Nd, Hf, Pb, etc.).

The MORB source is therefore depleted in the more incompatible elements both in the short and long term (Figs. 2.7 and 8.7). The oceanic lithospheric plate is isotopically indistinguishable from the

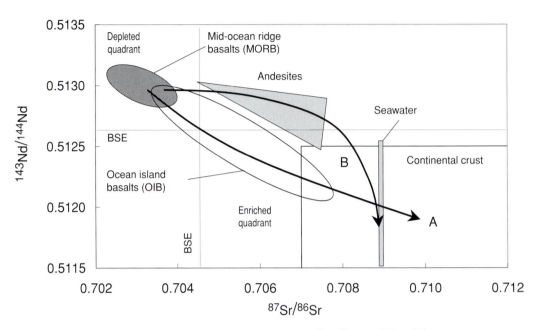

Figure 8.7: Relationship between $^{87}Sr/^{86}Sr$ and $^{143}Nd/^{144}Nd$ in the sources of oceanic basalts and andesites. The more incompatible character of Rb (compared with Sr) and of Nd (compared with Sm) during melting of mantle rocks explains the negative correlation of these ratios. MORB originates from a source that is more depleted in incompatible elements by successive melting of the mantle than the OIB source. This depletion is ancient (several billion years), as indicated by the composition of slowly accumulating radiogenic elements. Andesites derive from either hybridization of the mantle by the continental crust (A), or melting of the mantle contaminated by dehydration fluids rich in seawater (B). BSE stands for Bulk Silicate Earth.

asthenosphere, and so must be most of the upper mantle. Since granites, a major constituent of continental crust, are enriched in both the long and short term, a major concept has emerged according to which depletion of the upper mantle reflects the extraction of continental crust throughout the geological times. In contrast, the OIB source is generally depleted in the long term (as indicated by radiogenic isotopes) but enriched in the short term (as shown by their incompatible element distributions). A popular idea is that the OIB source has been enriched before melting by fluid and magma percolation reacting with the rock matrix: this process is known as metasomatism. In most cases, however, modern models of melting make this *ad hoc* interpretation rather unnecessary.

The orogenic magmas arising from the subduction zones, such as andesites, have a more complex history. If they are produced at a continental margin, such as the Andes, they clearly exhibit interaction between mafic magmas from the mantle and anatectic melts from the continental crust.

If they are produced in oceanic convergence zones, the result is more surprising: the isotopic composition of their Nd is very similar to that of the Nd of descending lithospheric plates. By contrast, their $^{87}Sr/^{86}Sr$ ratio is particularly radiogenic, suggesting that marine Sr is present in the mantle source of these magmas. This is often taken as evidence that dehydration of sediments and altered basalts of the downgoing plate produces fluids which trigger melting in the overlying mantle wedge. Andesites are often seen as the resulting melts. This interpretation is confirmed by higher $\delta^{18}O$ values (>7 per mil) than for the average of the ordinary mantle (≈ 5.5 per mil), therefore requiring the presence in their source material of rock components that have been subjected to weathering or alteration at low temperature. Andesites are richer in Si than basalts, which is consistent with the petrological concept stated earlier that water increases the Si content of melts. A number of trace elements exhibit abundance levels typical of such an environment: in particular the extreme depletion in Nb and Ta, which are thought to be locked in highly refractory minerals such as titanium oxide, in contrast to many incompatible elements (Rb, K, Ba, Sr, Pb, La, Th, etc.), which are returned toward the surface by dehydration of the plates. Petrology indicates that the origin of andesites cannot be ascribed to simple melting of the subducted crust as this process would produce lava far too rich in silica. This liquid is a special kind of granitic melt known as dacite. It can therefore be imagined that these liquids released by melting of the hydrated ocean crust react with the mantle located above the plate. Reaction between the mantle and the felsic liquids generated by hydrated fusion of the plate modifies the mantle overlying subduction zones: andesites are thought to derive from the melting of this anomalous mantle. The most popular interpretation is that andesites are melts formed upon invasion of the mantle wedge overlying the subduction zone by dehydration fluids.

Ocean island basalts (OIB) pose an enormous challenge to geochemists. The complexity of their isotopic characteristics has caused lasting confusion, which is only now beginning to clear. First, OIBs are isotopically heterogeneous. In the range of radiogenic isotopic ratios, e.g. for $^{87}Sr/^{86}Sr$, $^{143}Nd/^{144}Nd$, $^{206}Pb/^{204}Pb$, etc., these heterogeneities can only be accounted for by the mixture of a minimum number of isotopic "components." These are thought to correspond to different mantle sources, which each represents a different geodynamic object. The components are normally referred to by their acronyms: DM (depleted mantle) stands for a depleted source similar to the MORB source and to the asthenosphere found beneath the mid-ocean ridges; the HIMU component is characterized by particularly radiogenic lead (high μ, with $\mu = {}^{238}U/^{204}Pb$), a character that may be inherited from remelting of ancient ocean crust; EM (enriched mantle) stands for a mantle enriched

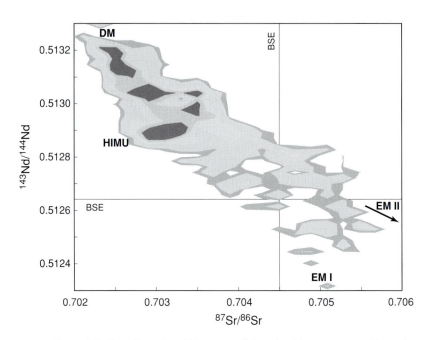

Figure 8.8: Two-dimensional histogram of Sr and Nd isotope compositions of oceanic basalts and the distribution of the four major mantle components (end-members): depleted mantle (DM, prevalent in MORB), high-^{238}U/^{204}Pb (HIMU) mantle (probably recycled oceanic crust), enriched mantle (EM) of type I (uncertain), and type II (recycled terrigenous sediments). BSE stands for the Bulk Silicate Earth.

in incompatible elements, either by incorporation of sediments (EM II) or by invasion of the mantle by mineralized deep fluids (EM I). Oceanic basalts normally occupy a specific range of isotope ratios, involving different radiogenic isotopes such as ^{87}Sr, ^{206}Pb, ^{207}Pb, ^{208}Pb, etc. (Figs. 8.8 and 8.9). Such chemical complexity is also found in the relative distribution of trace elements, the most incompatible of which are normally greatly enriched in OIB. This "component soup" has unfortunately barely furthered our understanding of oceanic basalt genesis.

With the new emphasis of some isotopic systems since the mid-1990s, notably the oxygen isotopes and the ^{187}Re–^{187}Os system, the complex geochemical landscape of the source of OIBs is coming into sharper focus. Isotopic variations of oxygen indicate that mantle material that had experienced low-temperature seafloor alteration is involved in the source of the Hawaiian basalts. Isotopic compositions of osmium and hafnium show that we find all the components of the ancient oceanic lithosphere recycled, from pelagic sediments (deep-sea clays), with high ^{176}Lu/^{177}Hf ratios, through the basaltic part of the weathered basalt crust, with high δ^{18}O and ^{187}Re/^{188}Os, to cumulate rocks (gabbro) of the lower oceanic crust with their low δ^{18}O and ^{187}Re/^{188}Os

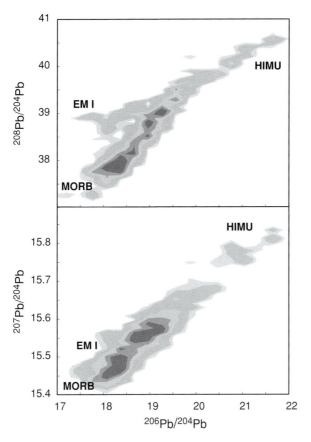

Figure 8.9: Two-dimensional histogram of Pb isotope compositions of oceanic basalts showing the high $^{238}U/^{204}Pb$ ratio of the mantle source of HIMU basalts (St Helena, Mangaia).

(Fig. 8.10). It is becoming ever clearer that the process forming most of the OIB source is recycling of old oceanic crust, possibly after melting or devolatilization of fluids in the course of plate subduction.

However, a mystery remains, that of isotopic geochemistry of the inert gases in the various types of basalt. Typical ridge basalts have $^3He/^4He$ ratios four times lower than basalt of ocean islands like Hawaii or Iceland. Let us refer back to (3.26), which we rewrite as:

$$\frac{d}{dt}\left(\frac{^4He}{^3He}\right)_t = \lambda_{238U}\left(\frac{^{238}U}{^3He}\right)_t\left[8+7\frac{\lambda_{235U}}{\lambda_{238U}}\left(\frac{^{235}U}{^{238}U}\right)_t + 6\frac{\lambda_{232Th}}{\lambda_{238U}}\left(\frac{^{232}Th}{^{238}U}\right)_t\right]$$

$$(8.4)$$

We first observe that $^{235}U/^{238}U$ is constant at a given time (1/137.88 today) and that $^{232}Th/^{238}U$ varies within narrow limits in the crust and the mantle (≈ 4 today). For $t = T = 4.55$ Ga, we obtain the following approximation:

$$\left(\frac{^4He}{^3He}\right)_T = \left(\frac{^4He}{^3He}\right)_0 + 18.7\left(\frac{^{238}U}{^3He}\right)_T\left(e^{\lambda_{238U}T}-1\right) \qquad (8.5)$$

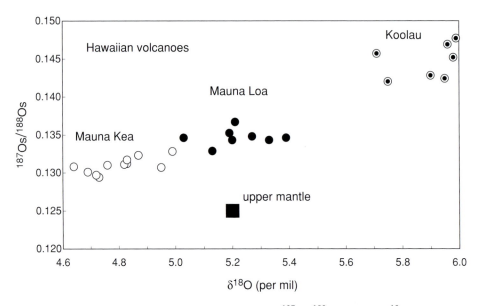

Figure 8.10: Relationship between the ^{187}Os/^{188}Os ratio and δ^{18}O of lava samples from three Hawaiian volcanoes. The variation in abundance in ^{18}O means that the source of these lavas contains a constituent that was processed in low-temperature surface conditions. Isotopic fractionation of Os requires the presence of a rock source that was once composed of ancient magmatic liquids and cumulates. The combination of these two characters suggests that these basalts remobilize ancient continental crust buried deep within the mantle. Other criteria preclude contamination by the present-day lithosphere (after Lassiter and Hauri, 1998).

We can sketch the evolution of the ^{3}He/^{4}He ratios in the mantle source of the ocean island and mid-ocean ridge basalts by assuming constant parent/daughter ratios (Fig. 8.11). Helium is distinctly more radiogenic in MORB than in OIB. The apparent ^{238}U/^{3}He ratio integrated over the Earth's history is therefore higher in the MORB source than in the OIB source. Given the much more volatile character of He compared with U and Th, it is inferred that parts of the OIB source have never been as intensely degassed as the MORB source, and so have never passed close to the surface. The presence of neon, whose isotopic composition is interpreted as signaling a gas component of solar origin in Hawaiian basalts, supports this argument. Such observations have long been considered proof of the primitive character of the mantle hot spots until the accumulation of isotopic data for non-volatile elements (Nd, Pb, O) made this hypothesis untenable. However, it is probable that very deep in the mantle there are still rocks that have remained intact since the Earth first formed, which we can illustrate with residence theory. Even if the mantle is well mixed for ^{3}He and assuming a helium residence time in the

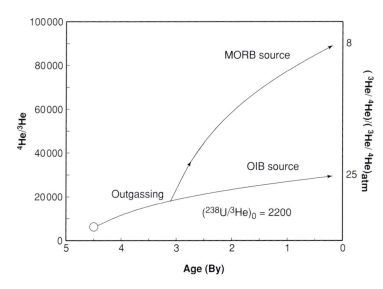

Figure 8.11: Helium isotope evolution of the mid-ocean ridge and ocean island basalt mantle sources, showing different $^{238}U/^{3}He$ ratios. Outgassing of helium upon melting produces mantle residue with higher $^{238}U/^{3}He$ ratios and therefore potentially more radiogenic helium, i.e. higher $^{4}He/^{3}He$ ratios, at a later stage. The more conventional notation of He isotopic compositions using atmospheric-He normalized $^{3}He/^{4}He$ ratios is shown on the right-hand side.

mantle of 1 Gy, we can use the equations developed in Appendix H to infer that a proportion of $e^{-4.5/1.0} \approx 1\%$ of the initial ^{3}He at the origin of the planet is still buried in the mantle. Certain rocks might act as reservoirs for this "primitive" gas whose isotopic signature is transferred, perhaps by diffusion, to old lithospheric debris accumulating in the mantle.

8.3 Mantle convection

Few problems have divided the Earth science community as much as the question of mantle dynamics. While all concur that convection really does occur, there is no consensus about the number of layers existing within the mantle and the way they contribute to its overall motion.

The idea that signs of recycling of ancient subducted tectonic plates could be found at the surface in the source of MORBs or OIBs arose in the late 1970s (Fig. 8.12, top). It is now envisaged that ancient oceanic crust, transformed into pyroxenite by pressure, supplies a large part of the fusible material that gives rise to OIB. New OIB can be made out of old MORB! In the model of whole-mantle convection, the plates sink to the base of the mantle. The intense flow of heat at the core–mantle interface causes gravitational instabilities and after one or two billion years the deep material rises rapidly in the form of plumes (blobs). Isotope and trace element data were persuasive evidence for this concept in the early 1980s, but four factors soon came to oppose the idea of whole-mantle convection based on simple export of lithospheric plates to the lower mantle:

- The rate at which material is exchanged within the mantle at the present-day rate of plate tectonics would eradicate any differences between the lower and

Figure 8.12: Two models of convection within the mantle. Top: whole-mantle convection. Subduction of lithospheric plates and rise of plumes form the convective system. Bottom: two-layer convection separated by the 660 km transition zone (notice the option of thermal coupling in the transition zone: upper mantle plumes are localized above the hot ascending zones of the lower mantle).

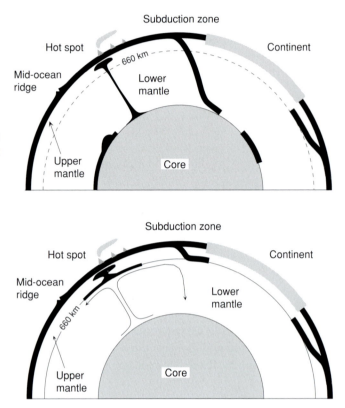

upper mantle in a very short time. This can be shown by replacing in (5.17) the values for residence time obtained from the mass of the two parts of the mantle and the flow of plates across the transition zone, to give a chemical relaxation time of 200–400 million years. Under these conditions, how could the geochemical identity and, above all, the isotopic identity of MORB be conserved relative to OIB?

- It is virtually impossible to find persuasive evidence of deep seismic activity that could indicate plate penetration into the lower mantle.
- With a mantle homogenized by convection, the radioactive element contents of the total mantle are equal to those of the upper mantle. So we can calculate the ratio between the heat produced by the Earth's radioactivity and that (43 TW) given off from the Earth's surface. This ratio, the Urey ratio, is then much smaller than unity. This would imply that the Earth still contains a considerable amount of the initial heat derived from the gravitational energy of accretion and core segregation locked at depth. If this condition is calculated back 4.5 billion years to the time our planet formed, it leads to an absolutely infernal temperature of formation, much higher than the vaporization temperatures of silicates.
- The argument from radiogenic argon has played a very significant role. The atmosphere contains a large quantity of radiogenic argon ^{40}Ar. If the total

quantity of ^{40}Ar produced by decay of terrestrial ^{40}K is estimated, and if allowance is made for the very low levels of this gas in the degassed upper mantle, an inventory of it for the planet can be drawn up. We quickly reach the conclusion that nearly 50% of the mantle was never degassed and so has never risen near the surface beneath ridges.

These powerful arguments are the basis for the canonical two-layer convection model (lower panel of Fig. 8.12). The upper and lower mantle are separated by the transition zone at 660 km and convect separately. Plates are recycled within the upper mantle and hot spots and their basalts (OIB) rise from the transition zone. The lower mantle is of virtually primeval composition and has very little involvement, supplying only argon and helium that is found in OIB.

A number of observations cloud this ideal picture. Except for rare gases, there is little evidence that any such primitive mantle contributes even in slight traces to the genesis of MORB or OIB. The second is that the terrestrial balance of some elements or isotopic systems is not at all consistent with the idea of a primitive lower mantle: Nb and Ta are depleted in both the continental crust and the MORB source with respect to similarly incompatible elements, the U–Pb, Re–Os, and Lu–Hf isotopic systems are not complete when known reservoirs are added up. The third observation was decisive. Modern high-resolution seismic tomography frequently shows lithospheric plates crossing the boundary between the upper and lower mantle. The canonical two-layer convection model of the 1980s with separation at the 660 km transition zone is no longer viable.

What is the current situation? If the 660 km transition zone is not taken to be an effective barrier, subduction zones must act as filters. As the material enriched in lithophile elements, which makes up the continental crust, is extracted at subduction zones, the material penetrating into the lower mantle must be greatly depleted. Moreover, the subducted oceanic crust transforms under pressure into eclogite, a very dense rock compared with the remainder of the oceanic lithosphere. It will tend to become delaminated and to accumulate by gravity at the base of the mantle and to build up an important store of fusible elements. It is also probable that convection has not fully homogenized the initial mantle of the planet and that, as will be discussed later, rocks inherited from the earliest time, may have survived convective erosion at very great depths ($>$1500 km). A combination of these two effects would justify the presence of U, Th, and K at depth and a reasonable Urey ratio.

8.4 The growth of continental crust

The extraction of the substance of continents from the mantle is an essential process in the evolution of the planet. It concentrates major

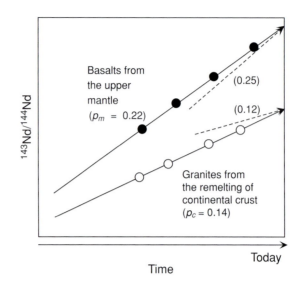

Figure 8.13: Isotopic evolution of Nd over time. The solid lines show the evolution of the $^{143}Nd/^{144}Nd$ ratio in mantle (basalt) and crust (granite) samples of different ages, the ratio being evaluated for the time the rocks were formed. The slopes of these two straight lines yield apparent $^{147}Sm/^{144}Nd$ ratios of the two geologic reservoirs. Conversely, the dashed lines show the changing $^{143}Nd/^{144}Nd$ ratio re-calculated over time using $^{147}Sm/^{144}Nd$ and $^{143}Nd/^{144}Nd$ ratios observed in samples of present-day mantle and crust. The difference in the slopes of the two curves over time is proof that the mantle and crust exchange material. It reflects the crustal growth phenomenon from the mantle and the recycling of continental crust back into the mantle.

elements, such as Si, Al, Na, P, and incompatible trace elements, such as Rb, Ba, the light rare-earths, and the radioactive elements, U, Th, and K, almost irreversibly near the surface. These elements share the characteristic of occurring in greater concentrations in melts than compatible elements. Proof that the continents are being permanently extracted from the mantle and that continental material is also being injected back into the mantle comes from isotopes. If we plot the evolution of certain isotopic ratios, and especially the $^{143}Nd/^{144}Nd$ ratio, against the age of formation of rocks, we generally obtain a graph as in Fig. 8.13. If the continents and the mantle had separated out since the beginning of geological time and had remained isolated from one another, the slope of the straight lines $p_c \approx 0.14$ and $p_m \approx 0.22$ would reflect the $^{147}Sm/^{144}Nd$ ratios of these reservoirs. This is not so. Typical $^{147}Sm/^{144}Nd$ ratios measured for continental rocks and the mantle are 0.12 and 0.25, respectively. This indicates that reservoirs corresponding to mantle and continents are open to mutual exchanges and have the effect of raising the apparent $^{147}Sm/^{144}Nd^{147}$ ratio of crustal rocks and lowering that of mantle rocks. We have put our finger on the reality of the growth of continents from the mantle and the recycling of continental crust by mantle convection.

Although there is apparent consensus that recycling of continental crust corresponds essentially to subduction of detrital sediments resting at the top of lithospheric plates, the inverse process of extraction of crustal material is not fully understood. As with convection, we can consider that there has been a canonical model of crustal growth based on the addition of andesite at converging plate boundaries. The Andean volcanoes, Mount Pinatubo or Mont Pelée, seem ideal agents for supplying mantle to the continents. This seems to be confirmed by the similarity of the chemical composition of lava from these volcanoes (including the famous Nb deficit) and the isotopic compositions of many elements, especially O, Sr, and Nd, to the composition of continental rocks.

Once again, superficial similarities mislead observers. A first worrying sign is the histogram of the age of formation of continental crust. Ages group around very well-defined peaks. There are "magic" ages for very rapid crustal growth around 600, 1100, 1800, 2700, or 3000 million years ago (Fig. 8.14). Geochronology indicates also that the time taken for these new continental expanses to form can be comparatively short. Millions of square kilometers of continent arose to form West Africa between 2170 and 2090 million years ago. Similarly, huge swaths of continental crust formed within a few tens of millions of years around 2700 Ma ago in the Superior Province of Canada. How can these characteristics be attributed to subduction, which is in essence a constant phenomenon in response to the formation of new oceanic crust, and therefore to the progressive

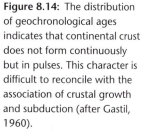

Figure 8.14: The distribution of geochronological ages indicates that continental crust does not form continuously but in pulses. This character is difficult to reconcile with the association of crustal growth and subduction (after Gastil, 1960).

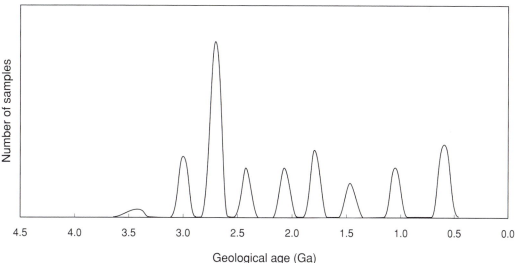

outflow of internal heat? We can, of course, imagine that cold oceanic lithosphere occasionally collapses into the deep mantle during brutal events referred to as an avalanche. The thermal effects and mechanical consequences of such a process appear too violent, though, compared with the relatively constant character of the geological record since the earliest times (lava chemistry, sediment mineralogy).

There is, nonetheless, a process of recurrent instabilities that can account for the episodic character of crustal growth. This is the eruption of plume heads, such as those at the origin of large igneous provinces (LIP) forming the traps of the Deccan in India or those of Ethiopia, and whose observable after-effects (plume stems) are the alignments of islands anchored to the hot spots of Hawaii or Réunion, among others. Other sudden submarine eruptions of very large size are known, such as the Ontong–Java plateau in the Pacific and the Caribbean plateau in the Atlantic. These plateaus currently form more than 10% of the ocean floor. Recent observations of Birimian formations in Africa (2100 million years) or Abitibi formations in Canada (2700 million years) show the remarkable geochemical similarity (Fig. 8.15) between continental protolith, i.e. the material extracted from the mantle and which coalesces to form continents, and the ocean plateaus. Although andesitic volcanic series exist in such deposits, they invariably overlie plateau material and so were formed subsequently. The subduction mechanism therefore builds up continental material from the oceanic plateau. When this material is thick enough to be buoyant and escape engulfment with the other oceanic crust, a new continent is formed. This construction is usually brief, taking less than 100 million years.

Figure 8.15: Distribution of concentration of rare-earths in the basalts of the Superior Province (Canada), the Birimian of West Africa, the pan-African of the Near East, and the Cretaceous Ontong–Java ocean plateau. This similarity suggests that continents formed from volcanic material similar to that issuing from plume heads (after Abouchami *et al.*, 1990 and Stein and Goldstein, 1996).

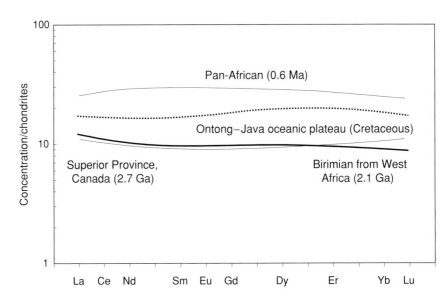

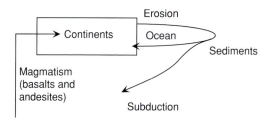

Figure 8.16: Crustal composition is the net result of an input – crustal growth out of magmas erupted from the mantle – and an ouput – erosion and subduction of sediments and oceanic crust. The composition of the continental crust does not therefore reflect that of any magmatism feeding it.

From a strictly geochemical point of view, the composition of the crust is a compromise between magmatic input from the mantle (basalt and andesite) and output through erosion. For all practical purposes, interaction with seawater can be ignored because the oceanic residence times of all elements are much shorter than the age of the continents. This compromise can be evaluated by using the method in Chapter 5. At first sight, the composition of the material that erosion removes from the continent, as shown by the composition of rivers and their suspended load, is very poor in silica, but rich in Mg and Ca. We cannot therefore discuss the origin of continents simply by comparing their composition with that of their protolith, i.e. the magmatic rock that may be their origin (Fig. 8.16). By way of analogy, because of respiration and removal of waste matter, the chemical composition of our bodies is not that of the food on our plates. Likewise continents must be seen as dynamic systems with both inputs and outputs. As hardly any significant differences in composition are observed between past and present crust, particularly in detrital sediments, it is considered that the system has reached a stationary state.

In fact, the assertion that present-day crust is indistinguishable from ancient crust applies only to its elemental content, whether of major elements, like Si and Mg, or of trace elements. When the isotopic compositions of Nd or Hf of basalts, either fresh or metamorphosed, at the time of formation are plotted, i.e. the composition of their source mantle, it can be seen that these compositions varied little, with slightly positive $\varepsilon_{Nd}(T)$ and $\varepsilon_{Hf}(T)$ values between 3.8 and 2.0 billion years (Fig. 8.17). This contrasts with virtually continuous growth of the radiogenic character of Nd and Hf from 2.0 billion years to the present. From the definition of notation ε given by (3.27) and evolution given by (3.23), the plot of Fig. 8.17 is largely identical to that of Fig. 3.11: for a closed system, the slope of the curve is proportional to the $^{147}Sm/^{144}Nd$ ratio of the source mantle. As the oldest known basaltic rocks (3.85 billion years from Greenland) already contain radiogenic Nd, it can be deduced that

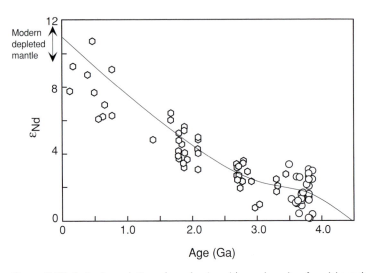

Figure 8.17: Isotopic evolution of neodymium (shown in units of ε_{Nd}) in rock from the mantle over geologic time. The slope of the curve of evolution is proportional to the Sm/Nd ratio. The existence of radiogenic Nd 3.8 Gy ago suggests that the Earth probably underwent intense fractionation of this ratio between 4.5 and 3.8 Gy ago (probably during its primordial differentiation). The Sm/Nd ratio remained low up to 2 Gy ago, then increased to its present-day value. Data compiled by Vervoort and Blichert-Toft (1999).

the ^{147}Sm/^{144}Nd ratio of their source mantle at the very beginning of the Earth's history (>4 billion years or Hadean) was higher than the chondritic ratio. Rocks rich in incompatible elements were therefore extracted from the Earth's mantle very early in its history, probably forming a continental proto-crust. This proto-crust was since reworked into more recent crust or engulfed by subduction into the lower mantle.

During the Archean and the early Proterozoic, the slope of the isotopic evolution curve for basalt Nd, and so the ^{147}Sm/^{144}Nd ratio of the mantle, remained fairly low and then, at around 2 billion years, progressively attained the modern value. Interpretation of this observation is complex and controversial because much continental crust seems to have been formed around 2.7 billion years ago. Two important points are involved in the explanation of this observation: (1) the Earth underwent substantial primordial differentiation that plate tectonics took one or two billion years to erase, and (2) because of the cooling of the Earth, the significance of violent convective instabilities in the form of hot spots diminished in favor of slower and more regular heat loss through ridges. Simply because heat production by radioactivity declined by a factor of three over the last four billion years, the plume flux must be far weaker today than in the Archean. This simple argument requires a reduction over the Earth's history of the juvenile magmatic contribution to crustal growth from hot spots. It is probable that if recycling of ancient crust is perceived as so important today, as for example in the Alps, the Himalayas, the Appalachians, or the Hercynian belt of Western Europe, it is because plumes have become much more infrequent.

References

Abouchami, W., Boher, M., Michard, A. and Albarède, F. (1990) A major 2.1 Ga event of mafic magmatism in West Africa on early stage of crustal accretion. *J. Geophys. Res.*, **95**, pp. 17 605–17 629.

Gastil, G. (1960) The distribution of mineral dates in space and time. *Amer. J. Sci.*, **258**, pp. 1–35.

Lassiter, J. C. and Hauri, E. H. (1998) Osmium-isotope variations in the Hawaiian lavas: evidence for recycled oceanic lithosphere in the Hawaiian plume. *Earth Planet. Sci. Letters*, **164**, pp. 483–496.

McBirney, A. R. (1992) *Igneous Petrology*. San Francisco, Freeman-Cooper.

Stein, M. and Goldstein, S. L. (1996) From plume head to continental lithosphere in the Arabian–Nobian shield. *Nature*, **382**, pp. 773–778.

Vervoort, J. D. and Blichert-Toft, J. (1999) Evolution of the depleted mantle: Hf isotope evidence from juvenile rocks through time. *Geochim. Cosmochim. Acta*, **63**, pp. 533–556.

Chapter 9
The Earth in the Solar System

The speed at which the universe is expanding as measured by the "red shift" of light from the stars indicates that the Big Bang occurred about 15 billion years ago. Because our star, the Sun, is a mere 4.5 billion-year old youngster, it is clear that the Solar System recycles chemical elements with a long history. These elements form in stars by a combination of processes of thermonuclear fusion, nucleon absorption, and radioactive decay. Who would think that our warm sunshine is actually the output of the largest nuclear reactor for light years around? The processes leading to the formation of elements are known collectively as nucleosynthesis.

9.1 The formation of elements

The most abundant element in the universe is hydrogen, an element formed by an electron orbiting around a single proton. Let us begin with the formation of a Sun-like star by gravitational collapse of an interstellar nebula, a cloud of gas and dust. The Sun itself, which makes up 98% of the total mass of the Solar System, consists of 71% H, 27% He, and 2% of heavier elements. The composition of the Solar System is given by spectroscopic analysis of sunlight, and it is this composition that will concern us here (Fig. 9.1). The accumulation of potential energy released by the collapse of the star's enormous mass raises the elements at its center to extremely high temperatures, typically several million K, until thermal agitation prevents further contraction. At these very high temperatures, the elements are totally stripped from their electrons, which leaves the Sun largely made of a dense gas of protons, electrons, and alpha particles. With the help of a tunnel effect, opposite to that referred to

165

Figure 9.1: Chemical composition of the Solar System standardized to a million atoms of Si. The main nucleosynthetic stages are marked above the curve (after Pagel, 1997). Notice the stability peaks (e.g. Fe, Pb) and the lower abundance of elements with odd atomic numbers compared with their even-numbered neighbors, whose nuclei are more stable. The deficit in the light atoms Li, Be, and B results from destruction of these elements in the stellar interiors.

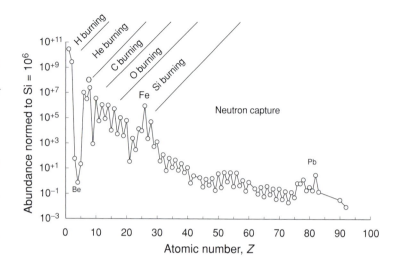

for the α decay process, the nuclei gain enough thermal energy to overcome the Coulomb repulsion barrier between them. The nuclei then fuse together and thermonuclear reactions begin, releasing gigantic energy reserves by conversion of mass. The star achieves steady state when the energy produced by nuclear reactions is balanced by the emission of neutrinos generated in the nuclear reactions and by electromagnetic radiation in outer space. At high pressure, however, the stellar gas is opaque. The highly energetic radiations produced in the stellar interior are therefore absorbed and further converted into thermal energy: they can only escape at a wavelength corresponding to the surface temperature of the star. The surface temperature of the Sun (6000 K) corresponds to radiation in the window of visible wavelengths.

Understanding these processes involves the concept of nucleon (protons and neutrons) bonding energy. In the fission of a uranium atom the total mass of the fragments, e.g. xenon and barium, is less than that of the initial atom. Conversely, in hydrogen atom fusion the mass of the resulting atom, e.g. lithium, is less than that of the initial hydrogen atoms. The extra mass is converted into thermal energy. Any reaction that releases energy is liable to occur spontaneously. If we plot the binding energy thus available per nucleon (Fig. 9.2) it can be seen that the maximum of the curve is located in the position of iron: if the activation barriers are crossed the fusion reactions of light elements and fission reactions of heavy elements will be spontaneous, while masses in the vicinity of iron will be stable.

Let us now examine the abundance distribution of elements in the Sun relative to their atomic number. A first observation is that elements with even atomic numbers are more abundant than those with odd atomic

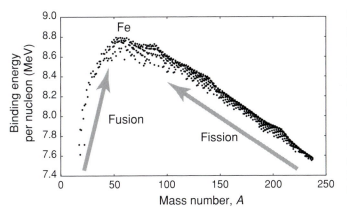

Figure 9.2: Binding energy per nucleon of different nuclides is maximum for iron (in millions of electron-volts). Fission of nuclides heavier than iron and fusion of nuclides lighter than iron are energy-producing reactions and, apart from their activation energy, occur spontaneously. Nucleosynthesis beyond iron will therefore consume energy.

numbers, reflecting the greater stability of nuclei in which the nucleons are paired. This characteristic is one justification of plots against a reference standard such as chondrites, which overcomes the odd–even effect. An irregular decline in abundance is observed with increasing mass number. In addition, abundance peaks occur at certain mass numbers such as that of iron (56), or for other nuclides with proton or neutron magic numbers of 2, 8, 20, 28, 50, 82, and 126. There is also a considerable apparent deficit in low-mass elements Li, Be, and B.

From hydrogen to iron, nuclear fusion within stars is the predominant process. The fusion of hydrogen into helium is the dominant "low" temperature (10^7–10^8 K) process and the one going on in the outer part of our Sun right now. The helium produced is burned at greater depth within the star and therefore at a higher temperature (10^8 K) to form C, N, and O, which in turn are raised to higher temperatures (10^8–10^9 K) and produce Mg, Al, Ca, and Si. Fusion stops at iron because production of heavier nuclides requires energy input. Li, Be, and B, for their part, are burned in the stellar interiors to form heavier elements and the small quantities of these elements observed in planetary material were produced in the interstellar medium upon breakup of heavy nuclei by cosmic rays.

At the very high temperatures (several billion K) resulting from gravitational collapse of the core of the more massive stars, explosive silicon burning produces the abundant elements of the iron group (Fe, Ni, Cr). For the elements beyond iron, processes of a different type are at work that involve the absorption of neutrons by nuclei. Three essential processes are the slow s process, the rapid r process and the p process, which stand for slow and rapid neutron absorption and for proton adsorption, respectively. Most elements are produced by a combination of processes although some of them are essentially pure, such

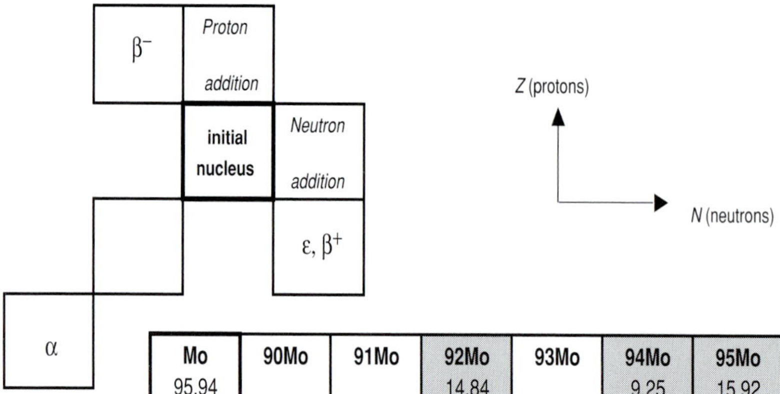

Decay scheme diagram (top left):

```
┌──────────┬──────────────┐
│   β⁻     │  Proton      │
│          │  addition    │
├──────────┼──────────────┤
│ initial  │  Neutron     │
│ nucleus  │  addition    │
├──────────┼──────────────┤
│          │   ε, β⁺      │
├──────────┼──────────────┘
│    α     │
└──────────┘
```

Z (protons) ↑   N (neutrons) →

Mo 95.94 Molybdenum			90Mo ε 5.7h	91Mo β⁺ 15.5m	92Mo 14.84	93Mo ε 3500a	94Mo 9.25	95Mo 15.92	96Mo 16.58
Nb 92.90638 Niobium			89Nb β⁺ 2.0h	90Nb β⁺ 14.6h	91Nb ε 700a	92Nb ε 3.6E7a	93Nb 100	94Nb β⁻ 2.0E4a	95Nb β⁻ 35.0
Zr 91 224 Zirconium	86Zr ε 16.5h	87Zr β⁺ 1.7h	88Zr ε 83.4d	89Zr e 3.27d	90Zr 51.5	91Zr 11.2	92Zr 17.2	93Zr β⁻ 1.5E6a	94Zr 17.4
Y 88.90585 Yttrium	85Y β⁺ 2.6h	86Y ε 14.7h	87Y ε 3.4d	88Y ε 107d	89Y 100	90Y β⁻ 2.7d	91Y β⁻ 58.5d	92Y β⁻ 3.5h	93Y β⁻ 10.2h
Sr 87.62 Strontium	84Sr 0.56	85Sr ε 64.8d	86Sr 9.86	87Sr 7.00	88Sr 82.58	89Sr β⁻ 50.5d	90Sr β⁻ 29.1a	91Sr β⁻ 9.5h	92Sr β⁻ 2.7h
Rb 85.4678 Rubidium	83Rb ε 86.2d	84Rb ε 32.9d	85Rb 72.16	86Rb β⁻ 18.7d	87Rb 27.84 β⁻ 4.8E10a	88Rb β⁻ 17.7m	89Rb β⁻ 15.4m	90Rb β⁻ 2.6m	91Rb

Figure 9.3: Excerpt from the chart of the nuclides with the number of neutrons on the x-axis and the number of protons on the y-axis. Stable nuclides are in gray. White on black are long-lived radioactive nuclides. White fields are short-lived radioactive nuclides. Each square shows the mass number, the isotopic abundance as a percentage, the decay process and the half life. The atomic mass of the element is shown at the start of the row. Notice the difference in dominant nuclear processes on either side of the stability zone: β⁺ and electron capture ε above the stability zone, and β⁻ below it.

as Ba (*s*) or Eu (*r*). The relatively minor *p* process produces nuclei rich in protons located above the zone of stable nuclides (gray in Fig. 9.3) by reaction of the nuclides with high gamma radiation emitted when supernovae explode, and these will not be discussed here. The *s* and *r* processes can be better understood by examining a segment of the chart of the nuclides (Fig. 9.3). It can be seen that the neutron-rich, unstable *s* and *r* nuclides (located to the right of the stability valley) decay by β⁻

emission, while the rare p unstable nuclides (left) decompose by electron capture or β^+ decay. This pattern of radioactive decay results in the preferential formation of nuclides in and around the valley of stability.

Neutrons are being produced permanently in stars by multiple reactions involving light elements. The s process starts beyond the peak of the iron group, i.e. beyond the top of the curve of the binding energy per nucleon (Fig. 9.2), with iron "seeds." It arises from high-temperature (300 million K) equilibrium between absorption of these neutrons by nuclei and radioactive β^- decay. The rate of change of the number n_i per unit volume of a stable nuclide containing i neutrons and which is bathed in a medium with a concentration of neutrons N_n agitated at the mean thermal velocity v_n is given by the equation:

$$\frac{\mathrm{d}n_i}{\mathrm{d}t} = -\sigma_i N_n v_n n_i + \sigma_{i-1} N_n v_n n_{i-1} \tag{9.1}$$

where σ_i and σ_{i-1} are coefficients with the dimension of a surface area, called neutron-capture cross-sections, that can be determined by laboratory experiments. The reader will recognize in this expression the neutron flux $N_n v_n$ through surface areas σ_i and σ_{i-1}. As the excess of neutrons over protons increases, the σ values decrease and neutron capture becomes less and less efficient, while the probability and the rate of β^- decay increase (Fig. 9.4). When the nuclide with neutron number i is radioactive, the previous equation must be changed to:

$$\frac{\mathrm{d}n_i}{\mathrm{d}t} = -\sigma_i N_n v_n n_i + \sigma_{i-1} N_n v_n n_{i-1} - \lambda_i n_i \tag{9.2}$$

where λ_i is its β^- decay constant. For a given nuclide, a steady state is reached where the input by neutron absorption on the progenitor equals the output by radioactive decay plus neutron absorption on the progeny, which, for stable isotopes, leads through (9.1) to the simple relationship:

$$\sigma_i n_i = \sigma_{i-1} n_{i-1} \tag{9.3}$$

This model provides a remarkably precise prediction of the abundance of a large number of elements and their isotopic proportions. An s pathway may be easily constructed, with its branchings and accumulation points, and the natural abundances of s nuclides can be successfully predicted as a function of the neutron flux. The abundance peaks (magic numbers) reported earlier correspond to extremely stable optimal filling of nuclear energy levels making the nuclei immune to neutron absorption. The s process stops at ^{209}Pb, when α decay becomes predominant and decomposes nuclei formed by the slow absorption of neutrons. Red giant stars are often thought to be home to s processes.

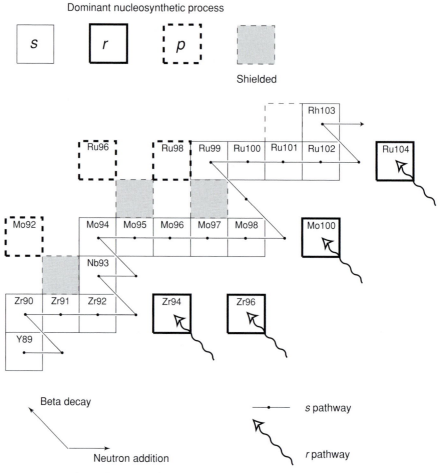

Figure 9.4: Example of nucleosynthesis by neutron capture for elements heavier than Fe. Slow *s* processes represent an equilibrium between neutron addition and β^- decay. Rapid (*r*) addition of neutrons in the expanding core of some stars produces neutron-rich isotopes by β^- decay. The *p* process yields proton-rich nuclei located above the valley of stability by reaction of the nuclides with high gamma radiation emitted by supernovae. Note the shielded isotopes between stable nuclides.

Other processes are required to explain many heavy nuclides, in particular those whose mass is greater than that of lead, such as U and Th. More neutrons are added to nuclei in the stellar cores, but also when supernovae explode, before β^- emission can occur. The *r* process of rapid absorption of neutrons by nuclei follows an *r* pathway quite remote from the valley of stability (Fig. 9.5) to which nuclides eventually return by β^- emission (Fig. 9.4). This fast track is "blocked" at the level of particularly stable nuclei unamenable to the addition of neutrons (magic

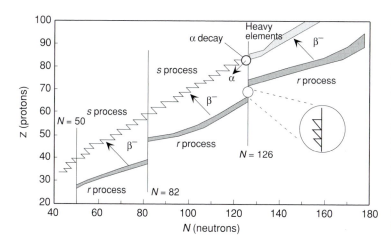

Figure 9.5: The origin of natural nuclides. The s process stops when α decay destroys the newly formed nuclei. When neutron absorption is too fast to allow loss of electrons (r process), it progresses to very stable nuclei (magic numbers) where it must wait for electron re-emission before it can go further (blown-up sawtooth). When the nuclei become too heavy, they are destroyed by fission.

numbers 50, 82, 126). Progression along the asymmetric sawtooth of magic numbers must wait at each notch for an electron to be lost by β^- emission before a new neutron can be absorbed. The process is stopped at the heavy masses when fission destroys nuclei faster than they can form. The abundance of r-process neutron-rich nuclides, as deduced by subtracting the contribution of the s process from the observed distribution of the elements in the universe, is well predicted by the expansion freeze-out of neutron-rich stars.

Overall, astrophysics has been very successful at predicting the abundance of elements and isotopes in the universe from the theory of nuclear physics and the experimental values of neutron cross-sections (the probability of neutron capture by a nuclide) and decay constants. It has been less successful at identifying the stellar sites at which this complex alchemy takes place.

9.2 The formation of the Solar System

The Solar System is composed of a medium-sized star – the Sun – and its retinue of planets. Some 99.9% of the mass of the Solar System is concentrated in the Sun and the dominant element is hydrogen. Our idea of how the Solar System formed goes back in its principle to Laplace. In a cloud of interstellar gas and dust (the Solar Nebula) swirling around its center of gravity, atoms, molecules, and particles fell to the equatorial plane. Thermal agitation caused countless collisions. Loss of momentum during these collisions caused material to spiral inward toward the Sun. Although most of the mass ended up at the heart of the system, the orbiting part that was to form the planets condensed into rock fragments (planetesimals) which coalesced through successive collisions.

The planets toward the inner part of the Solar System are more rocky because the high temperatures caused by intense radiation emanating from the Sun did not allow gases, such as water vapor and methane, to condense. It is commonly assumed that the "snow line," i.e. the imaginary limit of ice stability, was located somewhere near the asteroid belt between Mars and Jupiter. The outer planets such as Jupiter and Neptune formed in a colder environment and, in addition to their rocky core, also collected gaseous species such as hydrogen and methane. All the astrophysical models concur that the residual gas was blown away by the solar wind, the intense electromagnetic radiations (gamma rays, X-rays, protons, ^{3}He) and particle fluxes emitted by the Sun during its early gravitational collapse (T-Tauri phase), and this implies that Jupiter formed very rapidly, presumably within millions of years.

The numerous collisions between planetesimals increased the sizes of the resulting planetary bodies, reduced their number, and cleaned up their orbits. The gravitational pull of the massive outer planets perturbed their neighboring objects, which were either expelled from the Solar System or kept pouring down on the Sun and the inner planets. The outcome of these processes, which lasted less than 100 million years, was a small number of planets of similar size: Mercury, Venus, Earth, and Mars. An exception is the asteroid belt occupied by many small rocky bodies. The meteorites continually falling to Earth come from this zone and are the debris of asteroids whose orbit was strongly perturbed by the huge mass of neighboring Jupiter. Some of the meteorite parent bodies have been identified, such as the asteroid Vesta. Our planet is not the only one to be hit by such fragments. We know that impacts of such objects on the Moon and Mars have thrown out fragments now found as meteorites on Earth, and a constant albeit tenuous exchange of material between the various planets has been the rule over geological time.

A particularly significant collision led to the formation of the Moon. The strongest geochemical argument – and probably the strongest argument in general – for the common origin of the Earth and Moon is that the oxygen isotopes of rocks of both planets fall on the same mass fractionation line (Fig. 9.6), indicating thereby a common reservoir for this element. It is imaginable therefore that the impact of a body the size of Mars knocked off the material that now forms the Moon. At the time of impact, the Earth was still unfinished and its mass was only two-thirds of what it is today. The material of the impactor was highly refractory, greatly depleted in volatile substances, and intensely reduced.

The initial state of the terrestrial planets depended on their size and composition. The energy sources that may have contributed to heating the interior of these planets are well understood:

- Initial gravitational energy, which has markedly heated and almost certainly melted the external parts of the largest planets.

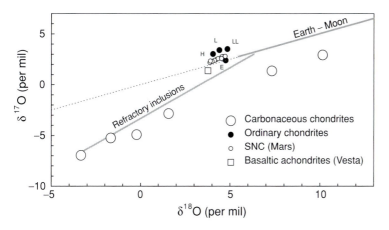

Figure 9.6: Isotopic composition of oxygen of different planetary bodies. The letters LL, L, H, and E next to the symbols of ordinary chondrites refer to chondrites with different iron contents and oxidation states. The refractory inclusions are globular assemblages of Ca-Al-rich minerals, such as spinel, anorthite, and melilite. They carry isotopic evidence that they existed very early in the history of the Solar System. Basaltic achondrites are igneous rocks chipped off from an asteroid thought to be Vesta. Samples of SNC (Martian) meteorites and ordinary carbonaceous chondrites do not lie on the straight line of isotopic fractionation of terrestrial and lunar samples. Note that the Earth cannot be made up of ordinary chondrites except from the E group (after Clayton and Mayeda, 1991).

- Gravitational energy of separation and crystallization of the iron and nickel core. This energy is enough to rise the Earth's temperature instantaneously by 2000 K.
- Radioactive heating by decay of short half-life nuclides such as ^{26}Al whose descendant ^{26}Mg has been identified in several planetary objects.

By analogy with Mars and the Moon, which have conserved rocks from their earliest age, it is thought that the Earth's core formed soon after the formation of the Earth itself. It is not known with certainty whether the Earth, like the Moon, went through a stage of near-general melting as what is termed a magma ocean, but it is very likely that it did. If so, it may be that the deep mantle still contains rock formed at that time and which could explain a number of isotopic anomalies, notably of inert gases, and account for certain properties of seismic waves.

9.3 Condensation of planetary material

The prediction of reasonable condensation sequences relevant to planetary accretion is one of the most successful contributions of thermodynamics to our understanding of the Solar System. The gravitational energy given off by the collapse of the solar nebula to form the Sun, and the radiative transfer of thermal energy thus produced, heated the proto-planetary material and vaporized it at temperatures in excess of 2000 K. As it cooled, the gas of atoms recombined and recondensed to form the solids, liquids, and gases observed at ambient temperature. Astrophysical models can give us temperature and pressure distributions throughout the nebula and their evolution over time. By applying elementary thermodynamic principles we should therefore be able to deduce the order of condensation of the various minerals involved in the formation of planetary bodies. Although it would be incorrect to assume that condensation is a well-understood equilibrium process, the insight into the diversity of planetary compositions and the mineralogy of planetary interiors provided by thermodynamic modelling is quite invaluable.

A first task is to draw up an inventory of the number of moles N of components (O, Al, Si, etc.) and potential species (CH_4, Al_2O_3, H_2O) regardless of their state, allowing the mass balance to be written, e.g. for silicon:

$$N_{Si} = N_{Si}(g) + N_{SiO}(g) + N_{Mg_2SiO_4}(s) + N_{MgSiO_3}(s) + \cdots \tag{9.4}$$

where g and s indicate whether the compounds are in a gaseous or solid state. As there are more unknowns than equations, the necessary number of mass action law equations must be completed. We can write gaseous equilibria of the type:

$$2H_2(g) + O_2(g) \Leftrightarrow 2H_2O(g) \tag{9.5}$$

and relate partial gas pressures by means of the mass action law:

$$\frac{P_{H_2O}^2}{P_{H_2}^2 P_{O_2}} = K(T, P) \tag{9.6}$$

where $K(T, P)$ is a known function of temperature and pressure. The condensation of planetary material produces different solid phases, which makes equilibrium calculations a tedious task. Let us write, for example, that magnesian olivine (forsterite Mg_2SiO_4) appears by condensation of Mg vapor, SiO, and water:

$$2H_2(g) + O_2(g) \Leftrightarrow 2H_2O(g) \tag{9.7}$$

$$2Mg(g) + SiO(g) + 3H_2O(g) \Leftrightarrow Mg_2SiO_4(s) + 3H_2(g) \tag{9.8}$$

By considering that olivine is nearly pure forsterite, the mass action law can be written:

$$\frac{P_{H_2}^3}{P_{Mg}^2 P_{SiO} P_{H_2O}^3} = K(T, P) \tag{9.9}$$

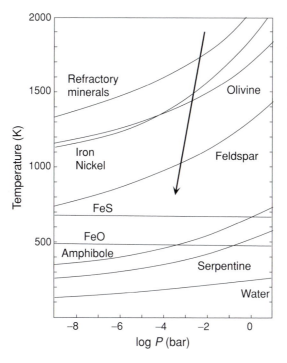

Figure 9.7: Condensation sequence of different minerals from the solar nebula calculated by adiabatic cooling (without heat exchange) of a gas of solar composition (after Lewis, 1995).

where $K(T, P)$ is the constant characteristic of this equilibrium. As most matter is initially formed from hydrogen it can be assumed that the partial hydrogen pressure P_{H_2} is equal to total pressure P_{tot} and that, applying Dalton's law, the pressure of each gaseous component is proportional to its molar proportion in the gas. Starting with a composition of solar gas, rather cumbersome calculations can resolve this complex system of equations with decreasing temperatures. The condensation sequence can be predicted from the most refractory minerals (such as melilite and perovskite of refractory inclusions of the famous Allende meteorite) through the common mantle minerals (olivine, pyroxene), the iron in the core, to the hydrated minerals (serpentine), and even water (Fig. 9.7). The mineralogy of each planet can be predicted fairly satisfactorily from its position in the proto-solar nebula and therefore from its temperature.

How effectively volatile elements are incorporated varies from planet to planet and depends on the distance from the Sun, which determines the accretion temperature, but also on the early history of each planet, in particular the massive impacts between planetesimals. If we compare a highly volatile lithophile element such as potassium with a very refractory lithophile element like uranium it can be seen that the K/U ratio, whose primordial value of 60 000 is given by carbonaceous chondrites, falls to 10 000 for the Earth (probably 20 000 for Mars), and to 3000 for the Moon. Our satellite is therefore particularly depleted in volatile elements. Reactions of the condensates with the residual ambient gas upon cooling of the nebula must also be considered as an efficient way of introducing volatiles into refractory material.

9.4 The composition of the Earth and its core, and the origin of seawater

The very difficult question of the Earth's composition has puzzled the geological community for over 40 years. Assembling the elemental inventory of the continental crust, the mantle, and the core is not an easy task since we have no robust estimate for the composition of the lower mantle and the core. We will see later that the Earth is essentially contemporaneous with the other planetary bodies of the Solar System for which we have samples: the Moon, the parent bodies of meteorites, and Mars. The major deep-seated fractionation processes, particularly the segregation of iron and nickel to form the core, which enable our planet to have a magnetic field, or the transfer of magmas to the surface, produced a strongly layered solid Earth. Closer to the surface, weathering, sedimentation, and metamorphism are other remarkable processes of chemical differentiation. The intense processing and the geochemical variety of terrestrial samples accessible to observation together with the existence of inaccessible domains in the deep Earth, such as the lower mantle and core, do not allow us to build up a verifiable picture of the mean composition of the Earth.

It is thought that the composition of the bulk Earth must resemble that of the least differentiated bodies of the Solar System. Chondrites are meteorites that have changed little since they first formed some 4.56 billion years ago. Their name derives from the presence of chondrules, droplets of liquids about a millimeter in diameter that crystallized rapidly in the nebula. The Sun and carbonaceous chondrites CI, of which one of the best known representatives is the Orgueil meteorite, have compositions that differ only by their contents of highly volatile elements such as hydrogen and nitrogen. The most common ordinary chondrites are depleted in gaseous elements and are more reduced than carbonaceous chondrites. Since chondrites come from the asteroid belt, well beyond the orbit of Mars, they did not suffer the same heating as the Earth, which was substantially closer to the radiation inferno of the nascent Sun. There is therefore no reason why the Earth should be identical in its composition to a particular type of chondrite, with the probable exception of the refractory elements. Chondrites represent an unlikely parent material for the Earth. The oxygen isotope composition of most chondrite classes, either carbonaceous or ordinary chondrites, is different from that of the

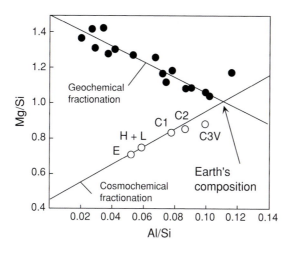

Figure 9.8: Evaluation of the composition of the primitive Earth by intersection of a trend for mantle rock and a trend for chondritic meteorites. The choice of refractory lithophile elements allows the share of the core and the crust to be ignored (after Jagoutz et al., 1979). E: enstatite chondrites; H, L: high-Fe and low-Fe chondrites, respectively; C: carbonaceous chondrites, with 1–3 defining the degree of metamorphism.

Earth–Moon system. A special category of meteorites, enstatite chondrites, could satisfy this criterion, but then we can infer from the Si content of the Earth's mantle that the core should contain 8–14 wt% Si, for which there is little supporting experimental evidence. In addition, the very low U and Th contents of enstatite chondrites is far too low to match the terrestrial surface heat flux.

The composition of the primitive mantle (BSE), i.e. the mantle prior to crust extraction, whatever the misgivings about its very existence as an extant geodynamic entity, can be determined through simple mass balance. It can be considered that the separation of the core had as its only effect to enrich the non-siderophile or chalcophile elements by a constant factor (ignoring the crust) in the silicate fraction. The compositions of meteorites and of the Earth's mantle are plotted on the graph of Mg/Si versus Al/Si (Fig. 9.8), along with that of fertile peridotites. The composition of these mantle samples is close to the composition assumed for the primitive mantle. The mantle and chondrites each define a trend. It is proposed that one trend corresponds to planetary petrological differentiation and the other to cosmochemical differentiation (selective vaporization) in the proto-planetary nebula. If the primitive material of the Earth belongs to the chondrite family it should lie at the intersection of these two alignments. Of course, we have no evidence that the alignment of mantle compositions amenable to sampling passes through the primitive mantle. If a significant reservoir of one of these elements is not correctly identified, the evaluation is incorrect. This is probably the case for the core: its density, estimated from the speeds of elastic waves, implies that it contains, in addition to Fe and Ni, light elements such as C and S, and maybe even a substantial fraction of Si, which biases the compositional models based on mass balance. In spite

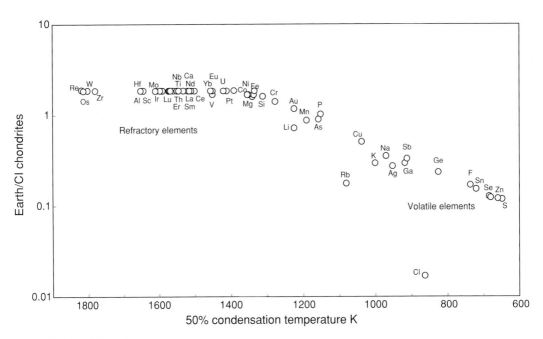

Figure 9.9: Depletion of the Bulk Earth with respect to the CI carbonaceous chondrites. For most elements except H and He, these chondrites have a composition similar to that of the Sun and therefore represent the most primitive solid material in the Solar System. The x-axis represents the temperature at which 50% of the initial element inventory has been taken up by solid phases in a standard condensation model. The Earth is strongly depleted in the most volatile elements such as K, Zn, S. Refractory elements such as Zr, La, Th, in contrast, have not been lost. They are slightly more abundant in the Earth than in CI because of volatile element depletion.

of these limitations, this method is still the most reasonable one to date.

Once the composition of the primitive mantle is obtained, the terrestrial concentration of the lithophile elements – which do not contribute to core composition – can be easily evaluated. It then becomes apparent (Fig. 9.9) that the Earth is depleted in the most volatile elements, such as K and Na. By comparison with ordinary chondrites, which also went through intense devolatilization, and using as a guide a scale of condensation temperatures, such as the temperatures obtained from the models described in the previous section, we can infer with some confidence the composition of our planet. To the best of our knowledge, the chondritic model works for the relative distribution of the refractory lithophile elements in the Earth. In particular, the chondritic concentrations of the ^{147}Sm–^{143}Nd and ^{176}Lu–^{176}Hf pairs of parent–daughter nuclides in chondrites are reliable indicators of planetary values and can be used as a robust reference for the Nd and Hf isotopic evolution of the Bulk Silicate Earth. This is in contrast to the ^{87}Rb–^{87}Sr, ^{238}U–^{204}Pb, and ^{187}Re–^{187}Os pairs in which at least one of the nuclides is either volatile (Rb, Pb) or siderophile (Os).

One aspect of the composition of the Earth and its mantle relates to the formation of its core. As indicated by the extinct radioactivity of ^{182}Hf (see below), the metallic core almost certainly formed within a few tens of My of the Earth's accretion. Core segregation is now seen as a runaway process releasing enormous amounts of gravitational energy, enough to melt the iron–nickel alloy and a substantial fraction of the mantle. It might be expected that during the segregation of the metallic core, highly siderophile elements strongly partitioned into liquid iron and were nearly completely removed from the mantle. Geochemical evidence tells us otherwise. Some moderately (Ni, Cr) to strongly (Pt, Re, Os) siderophile elements are far more abundant in the mantle than they would be if the mantle had been in contact with iron at the time it migrated into the core. Isotopic analyses for the parent–daughter ^{187}Re–^{187}Os pair indicate that the Re/Os abundance ratios in the mantle are chondritic, while the metal/silicate partition coefficients are very high ($>10\,000$) and different for both elements. A widely accepted explanation of this is that, shortly after the core segregated, the Earth was subjected to intense meteoritic bombardment and the highly siderophile elements observed in the mantle reflect this chondritic input, termed the late veneer. This process replenished mantle Re, Os, Ni, etc., in relative abundances identical to those found in meteorites.

A good model of the elemental composition of the Earth can provide some idea of its core composition. The properties of the Earth's magnetic field indicate that the core is composed of iron. Seismic shear waves do not propagate in the upper part of the core, which must therefore be liquid. The planetary inventory of nickel, the presence of nickel in iron meteorites, and metallurgical data, all indicate that the core is essentially an iron–nickel alloy (8% Ni). The densities obtained from wave propagation speeds also point to the presence of light elements such as oxygen, sulfur, and less probably magnesium and silicon. To evaluate core composition, McDonough (1999) suggests comparing the relative abundances of lithophile elements (elements that are a priori absent from the core) and siderophile elements of equivalent volatility (measured by the condensation temperature of these elements). For elements such as Fe, Ge, As, and S, the difference in abundance of lithophile and siderophile elements of equivalent volatility standardized to that of the Earth reveals a deficit represented by the elemental composition of the core. For all its elegance, this exercise depends on an assumption that, although reasonable, is difficult to test and leads to considerable uncertainties.

The issue of core composition is surprisingly germane to the question of where and when terrestrial seawater originated. It is reasonably well established by astrophysical models of star evolution that any original atmosphere was rapidly blown off from the Earth's surface by the intense radiations emitted by the nascent Sun. It seems reasonable to assume that primordial terrestrial water is best represented by water outgassed from oceanic basalts. This water was depleted in deuterium by about 80 per mil or more with respect to SMOW. Seawater therefore has a separate origin that some scientists do not hesitate to attribute to the outer Solar System. Comet showers are suspected to have been frequent during some particular geological periods and were considered first. Comets are celestial bodies originating in the outermost Solar System (the Kuiper belt) whose highly elliptical orbits bring them close to the Earth. They are made of abundant ice surrounding a small rocky nucleus. The measurement of the D/H ratio of Halley's comet by the Giotto probe revealed a D/H ratio twice that of seawater, therefore limiting the input of cometary water to less than 15% of the ocean. Alternatively, water may have been added to the Earth as frost from the abundant planetary bodies that, right after the Solar System formed, were cruising in the outskirts of Jupiter. These planetesimals were ejected very early from their orbit by the gravitational pull of this planet, which acted just like a giant slingshot. In either case, we would expect the terrestrial abundances of heavy rare gases, such as xenon, to be higher than they actually are. A mixed source, mantle outgassing, plus a contribution from distant planetesimals and comets, would probably account reasonably well for these observations.

Convection in the core powered by gravitational energy and radioactive decay sustains the Earth's magnetic field. Geophysicists working on the dynamo theory would like, however, to see core convection powered by more radioactive decay than is currently accepted by geochemists. Experiments of elemental partitioning between pure molten iron and silicate have so far been unable to incorporate substantial amounts of K into the metal. The class of iron meteorites that are believed to represent the core of proto-planets best are devoid of potassium. The two points of view could be reconciled if the high abundance of certain light elements such as Si, S, or O, which the velocity of seismic waves in the core requires, increased K solubility in molten iron alloy compared with that of pure iron.

9.5 The age of the Earth

Contrary to an often-heard assertion, the Earth itself has never been dated as there is no unmodified, inherited sample from the earliest age of our planet. In an atmosphere rife with religious prejudice, the age of the Earth was one of the great debates of the 19th and 20th centuries between geologists, who claimed much time was required for sedimentary strata to be deposited and organisms to evolve, and physicists. Using the theory of heat conduction, Lord Kelvin calculated an age of the Earth of 94 million years. One of the more imaginative attempts at dating the Earth was made in 1899 by the British astronomer Joly who calculated the time required for an ocean initially made of fresh water to become as saline as it is today if its only salt input was from rivers. This age τ can be calculated as the ratio of the mass of chlorine of seawater to the flux of chlorine

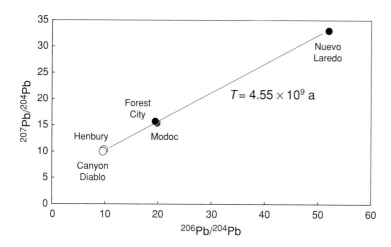

Figure 9.10: Patterson's geochron is in fact an isochron (^{206}Pb/^{204}Pb, ^{207}Pb/^{204}Pb) established from three chondritic meteorites (solid circles) and iron sulfide of two iron meteorites (open circles). The age established is therefore that of the Solar System.

from rivers, which using the data from Appendices A and G gives:

$$\tau = \frac{(1.4 \times 10^{21}) \times (1.89 \times 10^{-2})}{1.035 \times (3.6 \times 10^{16}) \times (7.8 \times 10^{-6}/1)} = 91 \times 10^{6} \; y$$

where allowance is made for the density of seawater (1.035) and of rivers. What we are really seeing here is the residence time of chlorine in seawater. In addition to the unsupported assumption of an initially fresh-water ocean, the theory described in Chapter 5 states that over the course of geological time the ocean reached the steady state for salt concentration. It therefore lost any memory of its origins and no significant age can be deduced by such a computation.

The existence of ages of more than three billion years was known to the earliest geochronologists, namely those who utilized lead isotopes. It became clear that Kelvin's calculations overlooked some fundamental processes. We now know that thermal evolution models of the Earth must include internal heating by radioactive decay of U, Th, and K, and allow for mantle convection. It was only in 1954, however, that Clair Patterson produced the first meteorite isochron on which he aligned chondrites and iron meteorites (Fig. 9.10). The age of 4.55 Ga indicated by this isochron has been challenged many times, but to no avail. Troilite (FeS) from one of these meteorites, Canyon Diablo, contains absolutely no uranium and therefore yields the initial isotopic composition of lead in the Solar System. The ca. 4.54 Ga age of the Earth has been repeatedly confirmed by all known chronometers, which in itself is startling corroboration of the validity of the principles of radiometric dating.

The oldest crustal segment on Earth (Isua, Greenland) is 3.85 Ga old. Very recently, zircons older than 4.4 Ga have been identified in an Australian sandstone, which requires the early existance of a granitic crust. Very ancient ages, of more than 4.4 Ga, have also been found for rocks of the Moon and Mars.

Patterson's isochron has improperly been called the geochron, although almost all the measurements of compositions of terrestrial samples lie not on the isochron but to its right. The essential argument for dating the Earth is, in fact, an indirect one relying on the method of extinct radioactivities, which provide a precise but relative chronology between planetary objects. The question then becomes "how much younger than primitive meteorites could the Earth be?" Let us establish the principles of the extinct radioactivity using the oldest known extinct chronometer, ^{129}I–^{129}Xe. Natural iodine currently has only one stable isotope ^{127}I, but at the time the solar nebula became isolated from the nucleosynthetic processes, it had another radioactive isotope of mass 129. ^{129}I has a decay half-life of 17 My and by now has completely decayed into its daughter isotope ^{129}Xe following the usual equation:

$$\left(\frac{^{129}\mathrm{I}}{^{127}\mathrm{I}}\right)_t = \left(\frac{^{129}\mathrm{I}}{^{127}\mathrm{I}}\right)_0 e^{-\lambda_{129\mathrm{I}}t} \tag{9.10}$$

In this equation, t is time counted from the origin, an arbitrary point after which the system can be considered closed (probably the end of nucleosynthesis), and $\lambda_{129\mathrm{I}}$ is the decay constant of ^{129}I. Let us imagine that minerals formed at time t when the nebula condensed. These minerals captured the two iodine isotopes, but at the moment of measuring (today) all of the ^{129}I has decayed to ^{129}Xe. The evolution of the ratio of a radiogenic ^{129}Xe isotope to a stable ^{132}Xe isotope can be represented by one of the standard equations for radioactivity:

$$\left(\frac{^{129}\mathrm{Xe}}{^{132}\mathrm{Xe}}\right)_{sample,t} = \left(\frac{^{129}\mathrm{Xe}}{^{132}\mathrm{Xe}}\right)_{sample,0} + \left(\frac{^{129}\mathrm{I}}{^{132}\mathrm{Xe}}\right)_{sample,0}\left(1 - e^{-\lambda_{129\mathrm{I}}t}\right) \tag{9.11}$$

Figure 9.11 shows changes in the ^{129}Xe/^{132}Xe isotope ratio in the different bodies that separated out from the solar nebula with different I/Xe

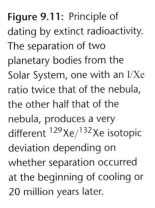

Figure 9.11: Principle of dating by extinct radioactivity. The separation of two planetary bodies from the Solar System, one with an I/Xe ratio twice that of the nebula, the other half that of the nebula, produces a very different ^{129}Xe/^{132}Xe isotopic deviation depending on whether separation occurred at the beginning of cooling or 20 million years later.

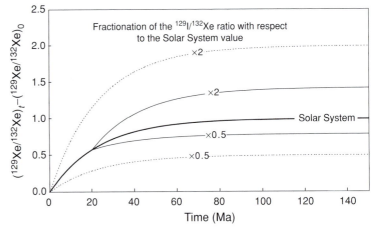

ratios. This representation shows the timekeeping virtues of the system we will now describe. The quantity of ^{129}Xe found today in the mineral is the sum of what was present at the moment it was trapped and of what was produced by the decay of ^{129}I:

$$\left(^{129}\text{Xe}\right)_{\text{sp},\infty} = \left(^{129}\text{Xe}\right)_{\text{sp},t} + \left(^{129}\text{I}\right)_{\text{sp},t} \tag{9.12}$$

where the sp subscript indicates the sample. This equation can be divided by the number of ^{132}Xe atoms, giving:

$$\left(\frac{^{129}\text{Xe}}{^{132}\text{Xe}}\right)_{\text{sp},\infty} = \left(\frac{^{129}\text{Xe}}{^{132}\text{Xe}}\right)_{\text{sn},t} + \left(\frac{^{129}\text{I}}{^{127}\text{I}}\right)_{\text{sp},t} \left(\frac{^{127}\text{I}}{^{132}\text{Xe}}\right)_{\text{sp},\infty} \tag{9.13}$$

In this equation we show that the planetary body from which a sample is analyzed condensed at t with the same isotopic composition of iodine and xenon as the solar nebula (sn) that engendered it. The symbol ∞ stands for a measurement at the present time, well after all the ^{129}I nuclides have decayed away. We also introduce the assumption that the proportions of the stable isotope ^{127}I remained unchanged since the formation of the sample. If the present-day isotopic composition of a representative sample of the Solar System, such as an unfractionated meteorite, can be analyzed we can write a similar equation for it and combine them as:

$$\left(\frac{^{129}\text{I}}{^{127}\text{I}}\right)_{\text{sn},t} = \frac{\left(^{129}\text{Xe}/^{132}\text{Xe}\right)_{\text{sp},\infty} - \left(^{129}\text{Xe}/^{132}\text{Xe}\right)_{\text{sn},\infty}}{\left(^{127}\text{I}/^{132}\text{Xe}\right)_{\text{sp},\infty} - \left(^{127}\text{I}/^{132}\text{Xe}\right)_{\text{sn},\infty}} \tag{9.14}$$

It can be seen then that the measurement of ordinary iodine and xenon in the samples and the measurement of the isotopic composition of xenon are sufficient to give the $(^{129}\text{I}/^{127}\text{I})_{\text{sn},t}$ ratio. The $(^{129}\text{I}/^{127}\text{I})_0$ isotope ratio is not well known, but by taking a particular meteorite as a reference, in this case the Bjurböle chondrite, (9.10) can give a relative age of formation of all the objects of the Solar System on which measurements are possible.

We have great difficulty, of course, in evaluating ^{129}I and ^{132}Xe for the whole Earth and therefore this method, given for the sake of illustration, is not applicable as it stands. However, differences of $(^{129}\text{Xe}/^{132}\text{Xe})_{\text{sp},\infty}$ are observed between the Earth's atmosphere, meteorites, ridge basalts, and deep carbonic gas wells. This clearly indicates that when the Earth formed from the solar nebula and when the atmosphere separated out from the mantle there was still sufficient ^{129}I around to create isotopic variations of xenon. Bearing in mind that the half-life $T^{1/2}$ of this nuclide is 17 My and that eight half-lives are enough to eliminate 99.9% of the atoms of a radioactive isotope, the formation of the Earth and the degassing of the atmosphere must have occurred within a few tens of millions of years after the end of nucleosynthesis. The very early formation of the atmosphere is confirmed by the extremely radiogenic character of the argon (high $^{40}Ar/^{36}Ar$ values) of the mantle and mid-ocean ridges. There are other isotopes derived from extinct radioactive nuclides that inform us of other events in the early Solar System, such as ^{26}Mg from ^{26}Al ($T^{1/2} = 0.7$ My), ^{53}Cr from ^{53}Mn ($T^{1/2} = 3.7$ My), and ^{142}Nd from ^{146}Sm ($T^{1/2} = 103$ My). The isotope 182 of tungsten also presents abundance variations attributable to decay of the ^{182}Hf nuclide, whose half-life is 9 My. In contrast with tungsten W, which is largely siderophile, Hf is lithophile: core formation is the prime suspect for W/Hf fractionation and we conclude that the core was also transformed in the 30 millions of years of the Earth's history. While lead isotopes also tell us that intense fractionation of terrestrial uranium (lithophile) and lead (chalcophile) attributable to the separation of the core occurred at a very early stage, the extinct radioactivity of ^{182}Hf allows us to zoom in on this period which so far appeared a priori inaccessible.

It should be recalled therefore that the Earth has not been dated directly but that chondrites have, and with great precision (4.56 billion years). The formation of the Earth, of the other bodies from which samples are available (Mars, the Moon, and some asteroids, such as Vesta) must have been completed within a time interval of 10–20 million years, while all of the essential early processes that affected them, such as the formation of their cores and atmospheres, may have lasted another 30–50 million years.

9.6 The Moon

The Apollo missions enabled the collection of several hundred kilograms of samples from six localities on the visible side of the Moon and a wealth of observations by the astronauts. Although oxygen isotopes tell us that the Earth and the Moon formed from the same mass of debris left by the collision of a body the size of Mars with the proto-Earth, our satellite has a very different history and geology from that of our planet. The Moon's radius is 1738 km and its gravity a mere 17% of that of the Earth; a consequence of this is that the Moon does not retain most gases and therefore has no atmosphere. Gravimetric and altimetric data sent back by the satellite Clementine tell us that the Moon has a crust some 60–100 km thick, much thicker than the Earth's continental crust. A metal core was probably present in the early history of the Moon, but the absence of any outer magnetic field shows that it has now solidified. Chemically, the Moon is depleted in volatile elements and is extremely reduced, so much so that metal iron is a common mineral in magmatic rocks. It is a dead body whose last volcanic eruptions occurred 3.2 billion years ago.

Lunar rocks are classified in three main groups: (1) intrusive rocks of the lunar highlands, especially anorthosite, with its high plagioclase

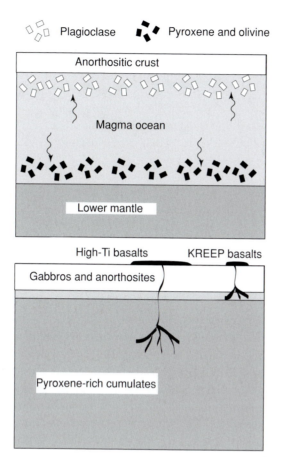

Plagioclase — Pyroxene and olivine

Anorthositic crust

Magma ocean

Lower mantle

High-Ti basalts   KREEP basalts

Gabbros and anorthosites

Pyroxene-rich cumulates

Figure 9.12: The magma ocean from which the lunar mantle and crust originated. Heat from the impact that separated the Earth and the Moon left the outer part of the Moon molten. Plagioclase feldspar floated to create the anorthositic crust (lunar highlands), while olivine and pyroxene settled out to form the mantle. KREEP basalts derived from the last residual liquid accumulated beneath the crust. Basalts from lunar maria derive from the subsequent re-melting of mantle cumulates.

content, and gabbro, which, in addition to plagioclase, contains olivine or pyroxene; (2) KREEP basalt, an acronym indicating that it is very rich in potassium, rare-earths, and phosphorus; and (3) lunar mare basalts. The intrusive rocks are 3.8–4.5 Ga old, KREEP basalts 3.6–3.8 Ga old, and mare basalts 3.2–3.9 Ga old. The feldspathic highlands dominate the far side. Very large impacts, such as the one that excavated the Mare Imbrium at around 3.8 billion years, formed a consolidated rubble known as breccia, which is particularly abundant in the samples brought back by the astronauts. Chemical maps produced by the satellites Clementine and Lunar Prospector show that the ejecta forming the wall of the Mare Imbrium are particularly rich in KREEP. The surface of the Moon is covered with regolith, a thick soil created by incessant meteoritic bombardment (gardening) of a body with no protective atmosphere.

The most popular model for the formation of lunar rocks is that of the magma ocean (Fig. 9.12). Observations on the first samples of breccia, brought back by Apollo 11, suggested that the relatively abundant anorthosite is an essential component of the surface while lunar mare basalt chemistry indicated a form of basalt–anorthosite

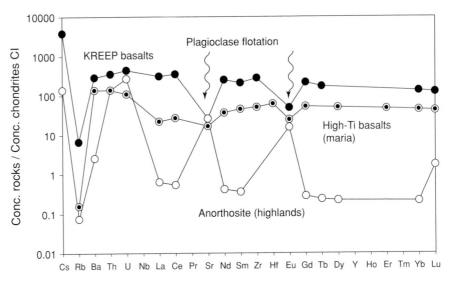

Figure 9.13: Trace elements
in lunar rocks. Anorthosite and
gabbro of the lunar highlands
exhibit surplus Eu and Sr from
the accumulation of
plagioclase by buoyancy (cf.
the partition coefficient of
Fig. 2.5). KREEP basalts exhibit
negative anomalies
complementary to those of
anorthosite. Lunar maria
basalt, being older, displays
similar negative anomalies
inherited from the source in
the mantle and depleted in
plagioclase.

complementarity. It is thought that after the impact that formed the
Moon some 4.56 billion years ago, its mantle was largely melted. As
this enormous mass of magma cooled, which may have taken several
hundred million years, all of its upper part was saturated in plagioclase,
olivine, and pyroxene. Plagioclase, being lighter than the liquid, floated
to form the highland crust, while the heavier olivine and pyroxene sedimented out, filling the magma ocean from the bottom upward. The last
residual liquids, particularly rich in incompatible elements (including K,
P, the rare-earths, and Ti), subsisted for several hundred million years
beneath the anorthositic crust. What arguments are there to support this
model? The first is the presence of a strong positive europium anomaly
in the anorthosite and other intrusive rocks, while the KREEP basalt
displays a very marked negative anomaly (Fig. 9.13): given the effective incorporation of Eu in plagioclase compared with other rare-earths
(Fig. 2.5), we are led to think that efficient segregation of plagioclase by
buoyancy from a magmatic liquid was at the origin of the intrusive rocks
of the lunar highlands. The KREEP basalts are formed out of a mantle
impregnated with residual liquids: the very high content of their mantle
source in heat-producing elements (U, Th, K) may actually have kept this
particular layer partially molten for billions of years. In the same way, the
unradiogenic character of the Nd of the anorthosite at the time it formed
contrasts with the radiogenic Nd of the basalts (Fig. 9.14). The inference
from this is of complementary Sm/Nd fractionation processes of the
early lunar mantle, which is further evidence of magmatic fractionation.

How did the basalts of the lunar maria form in this process? Their
relatively young ages exclude that they erupted out of the magma ocean

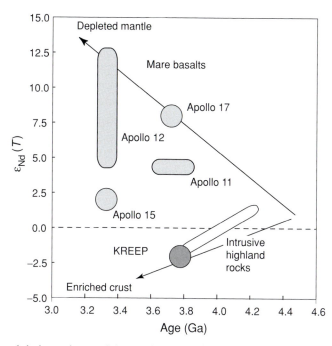

Figure 9.14: Evolution of the Sm–Nd isotope system on the Moon. Anorthosite of the lunar maria and KREEP basalts indicate geochemical "enrichment" (low Sm/Nd ratio and therefore unradiogenic Nd). By contrast, Nd of the source of lunar maria basalts is radiogenic, indicating that the mantle source is geochemically depleted, which is consistent with the hypothesis of pyroxene accumulation. Compare this with the evolution of the Earth's mantle and crust (Figs. 8.13 and 8.17).

system and their modest enrichment in incompatible elements indicates that they do not represent residual liquids. Nonetheless, these basalts exhibit a highly negative Eu anomaly indicating that their mantle source is a complement to the lunar highland crust. The most common interpretation is that they formed by re-melting at around 3.8–3.2 billion years of pyroxene-rich cumulates present in great quantities in the mantle and produced by crystallization of the magma ocean. The highly variable titanium contents of these basalts show that another mineral, ilmenite ($FeTiO_3$), was also present in very variable quantities at the time of melting. The sources of these lavas in the lunar mantle therefore represent cumulates of a magma ocean at relatively advanced stages of crystallization.

The essential feature of lunar mantle differentiation is the saturation in plagioclase of the magma ocean. On Earth, basaltic magmas can precipitate plagioclase down to about 30 km. On the Moon, where gravity is six times less, the equivalent pressure is only reached at about 180 km. It can be seen therefore that potential plagioclase production by lunar magmas is much more substantial than on Earth and that the potential production of anorthositic crust is much greater on the Moon.

9.7 Mars

Mars has a radius of 3386 km and its gravity is 38% of that of the Earth. Mars is a heavy planet for its size, which is usually taken as a sign of a high iron content. The planet retains, admittedly somewhat imperfectly, an atmosphere, with a pressure equivalent to 0.7% of the Earth's atmospheric pressure and which is capable of generating violent winds. The composition of the atmosphere, reported by the Viking landers, is dominated by CO_2. The orbital satellite MOLA told us that the thickness of the Martian crust varies from 40 to 80 km. There is a marked contrast between the two hemispheres: the southern hemisphere is composed of high volcanic plateaus, which are heavily cratered and therefore probably very old, dotted with volcanoes of fresh morphology and eroded by what are unquestionably fluviatile systems. The northern hemisphere is a vast flat depression increasingly interpreted as the floor of an ancient ocean. A polar ice cap composed mainly of carbonic ice forms and melts annually at high latitudes.

Although there have not yet been any missions to recover samples from Mars, the origin of more than 20 meteorites has been ascribed to Mars by powerful arguments. These meteorites are known as SNC meteorites (from the three group-leading meteorites Shergotty, Nakhla, and Chassigny). The arguments that they are samples from Mars are: (1) these meteorites form a homogeneous group in the ($\delta^{17}O$, $\delta^{18}O$) plot (Fig. 9.6); (2) the relative abundance of gases trapped in the Martian meteorite EETA79001 found in the Antarctic ice is identical to that of the atmosphere analyzed by the Viking probes (Fig. 9.15), and; (3) the relatively young ages (160–600 million years) found for most SNCs require the existence of a planet with recent volcanic activity,

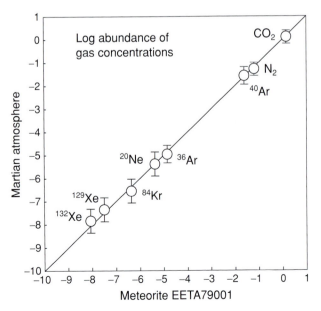

Figure 9.15: Gas in SNC (Martian) meteorites. The similarity of their gas content to that of the Martian atmosphere measured by the Viking probes (standardized to nitrogen) is a very forceful argument for SNCs originating from Mars.

while the ages of exposure to cosmic radiation determined from these meteorites indicate that they have not spent more than a few million years in interplanetary space and that they probably came from not very far away. Mars and its super-volcanoes, with their very well preserved morphology, such as Olympus Mons and its caldera with vertical cliffs several kilometers high, is the only planet in the inner Solar System that could be a possible candidate.

The SNCs are cumulate rocks (gabbro, peridotite) affected by the impact that allowed them to escape Martian gravity to the extent that the plagioclase has lost its crystalline structure and become amorphous. Little is certain from this blind sampling except that the SNCs must represent the Martian lithosphere. The isotopic compositions of Sr, Nd, and Hf reveal a very clear contrast between rocks from a source enriched in incompatible elements and others from a depleted source: we can contrast samples from the Martian "mantle" and samples from a Martian "crust." The existence of isotopic anomalies left by the extinct radioactive nuclides ^{146}Sm and ^{182}Hf and their variation from one meteorite to another further indicate that the Martian lithosphere is as old as the planet itself. Plate tectonics is therefore quite unlikely to have modelled the planet's surface. There are grounds for thinking that it is a very long time since Mars experienced any intense internal activity, with the possible exception of the few volcanoes visible at its surface, and that even its metal core is frozen as it does not generate a magnetic field.

However, it is generally accepted that Mars once had a denser atmosphere than today and even oceans. The presence of sulfates and of low-temperature carbonates in SNC meteorites are evidence of water circulation, corroborating satellite observations of hydrographic networks, probably of very great age. The existence of biological activity – a hypothesis given much media coverage by NASA thanks to photographs of minute elongated calcitic objects in the Martian meteorite ALH84001 discovered in Antarctica – encounters a degree of disbelief among the scientific community. The wider implications of the claim for society are such that the possibility must, of course, be further investigated.

References

Clayton, R. N., Mayeda, T. K., Goswami, J. N. and Olsen, E. J. (1991) Oxygen isotope studies of ordinary chondrites. *Geochim. Cosmochim. Acta*, **55**, pp. 2317–2337.

Jagoutz, E., Palme, H., Baddenhausen, H., Blum, K., Cendales, M., Dreibus, D., Spettel, B., Lorenz, V. and Wänke, H. (1979) The abundances of major, minor, and trace elements in the earth's mantle as derived from primitive ultramafic nodules. *Proc. Lunar Planet. Sci. Conf. 10*, pp. 2031–2050.

Lewis, J. S. (1995) *Physics and Chemistry of the Solar System*. San Diego, Academic Press, 537p.

McDonough, W. F. (1999) The Earth's core, in: C. P. Marshall and R. F. Fairbridge (Eds) *The Encyclopedia of Geochemistry*. Amsterdam, Kluwer, pp. 151–156.

Pagel, B. E. J. (1997) *Nucleosynthesis and Chemical Evolution of Galaxies*. Cambridge, Cambridge University Press. 329p.

Chapter 10
The geochemical behavior of selected elements

This chapter is intended to provide a geochemical overview of a number of important elements. The elements will be grouped according to mixed criteria, in particular their position in the periodic table and their geochemical properties. We will describe the major mineral phases that host these elements in the mantle and the crust, their properties in solution, and the processes by which they are transferred from any major reservoir (mantle, crust, ocean) to its neighbors. We will not reproduce here the terrestrial abundances, which can be found in Appendix A. We will nevertheless provide the reader with some important data. Condensation temperatures in the solar nebula (Wasson, 1985) define the volatile versus refractory character of the element. The solubility and complexation data in surface waters (Morel and Hering, 1993) and the residence times in seawater (Broecker and Peng, 1982) constrain the concentration level and speciation in natural waters at low temperature. Different parts of geochemical cycles may receive uneven attention. This inhomogeneous treatment reflects the power of geochemistry: different elements are used to trace different processes. The elements that essentially fractionate in atmospheric and oceanic processes (N, O, H) have not been dealt with, while only the long-term aspects of the carbon cycle have been considered.

10.1 Silicon

Most common form: Si^{4+}
Ionic radius: 0.26 Å
Stable isotopes: 28 (92.23%), 29 (4.67%), 30 (3.10%)

Atomic weight: 28.086
Condensation temperature: 1311 K
Dissociation of silicic acid in water:

$$H_2SiO_3 \Leftrightarrow HSiO_3^- + H^+ \ (-\log K = 9.86)$$

No significant complex in waters
Reactions limiting solubility in water:

$$SiO_2 + H_2O \Leftrightarrow H_2SiO_3 \ (-\log K = 4.0)(\text{quartz})$$
$$SiO_2 + H_2O \Leftrightarrow H_2SiO_3 \ (-\log K = 2.7)(\text{amorphous silica})$$

Residence time in seawater: 20 000 years

Silicon is, after oxygen and iron, the third most abundant element in the Earth. Although Si concentration in the core may reach several per cent, this ubiquitous element is essentially lithophile and refractory. It is the most abundant cation of the silicates that form the mantle at shallow depth (olivine, pyroxene, plagioclase, garnet), at intermediate depth (ringwoodite, majorite), and in the deep mantle (perovskite). It also forms the major igneous minerals in the crust of both mafic igneous rocks (olivine, pyroxene, amphibole, plagioclase) and felsic igneous rocks (quartz, feldspars). Silicon is a major constituent of clastic sediments (quartz and clay minerals) and makes up a substantial fraction of metamorphic minerals.

At pressures corresponding to the surface, the crust, and the upper mantle, Si is tetrahedrally coordinated and occupies the center of a four-oxygen tetrahedron. At very high pressure, Si can be accommodated in octahedral sites. Since melts from the mantle are systematically enriched in silicon with respect to the residue, this element may be classified as slightly incompatible. It forms stoichiometrically fixed minerals: the various forms of silica SiO_2 are stable at different pressures: quartz and amorphous silica are stable in the crust and the upper mantle, while coesite and stishovite are stable at increasingly higher pressures. A basalt typically contains 45–50 wt% SiO_2 and granitic melts 70 wt% SiO_2. Quartz is unstable in the presence of Mg-rich olivine. Silica solubility in hydrous fluids increases with temperature and with pH (see Chapter 7).

High-temperature fluids transport large quantities of SiO_2, which, upon cooling, form the ubiquitous quartz veins seen in metamorphic basements. Warm diagenetic fluids carry silica, which precipitates in the upper sedimentary layers: this is the origin of the familiar flintstones that grow on any seed (fossil, pebble) occurring in limestones. When the replacement of the initial sediment is total, a siliceous rock known as chert is obtained.

As a result of the very low solubility of silica at ambient temperature, quartz is essentially left untouched by erosion and forms the familiar sands that, upon diagenetic cementation, become sandstones. For the same reason, silica concentration in seawater is very low and, as a result of biological activity, the ocean is actually undersaturated in silica. SiO_2 is used by both phytoplankton (diatoms) and zooplankton (radiolara). Silica is therefore depleted in the upper layer of the oceans, while the falling debris re-dissolves at depth. Deep old waters are the most enriched in silica. Oceanic upwellings, such as the Atlantic Ocean off the coast of Morocco or the equatorial belt, bring up abundant sources of silica. Sediments from those areas are rich in silica deposited by falling organisms. Other localities, with warm hydrothermal springs on the seafloor, such as in the neighborhood of ridges and other volcanic areas, are also rich in siliceous sediments. The older waters from the Southern Oceans being rich in silica, it is not surprising to see a substantial fraction of the seafloor of the southern hemisphere covered with diatom-rich oozes.

10.2 Aluminum

Most common form: Al^{3+}
Ionic radius: 0.39 Å (tetrahedral) and 0.54 Å (octahedral)
Stable isotope: 27 (100%)
Atomic weight: 26.982
Condensation temperature: 1650 K
Complexes in water: hydroxides
Reactions limiting solubility in water:

$$\frac{1}{2}Al_2Si_2O_5(OH)_4 + \frac{7}{2}H_2O \Leftrightarrow H_4SiO_4 + Al(OH)_4^- + H^+ \ (-\log K = 19.5)$$

(kaolinite)

Residence time in seawater: 620 years

Aluminum is the sixth most abundant element in the Earth. Al is a highly refractory lithophile element. The radioactive isotope ^{26}Al quickly decayed into ^{26}Mg in the very first millions of years of the Solar System's evolution. It provided substantial heating to the early planets, and the isotopic composition of Mg is one of the most widely used chronometers based on extinct radioactivities. It is unlikely that Al enters in large proportions in the core, but Al is a major constituent of many major minerals at any depth in the mantle and in the crust. In the mantle, it enters plagioclase up to pressures of about 1 GPa, spinel to 2 GPa, and garnet beyond. At these high pressures, Al also enters clinopyroxene in large proportions: garnet and clinopyroxene dissolve into each other to form majorite, an essential mineral phase of the mantle above the 660 km discontinuity. At higher pressure, Al is hosted in a perovskite structure, but its precise behavior is still largely unknown. The major mineral that hosts Al in igneous rocks is feldspar: only plagioclase occurs in basalts, while plagioclase and alkali feldspar may occur together in felsic rocks. Biotite mica may occur in both types of rocks but normally accounts for a small part of the Al inventory. In sedimentary rocks, Al is hosted in clay minerals such as kaolinite and illite and, occasionally, in detrital feldspars. In metamorphic gneisses and schists, Al largely resides in feldspars and micas.

Aluminum can be tetrahedrally coordinated and in this coordination it replaces Si in the center of oxygen tetrahedra. It can also be octahedrally coordinated and form solid solutions with elements such as Ca, Mg, and Fe. During melting, the Al-rich minerals (feldspar, spinel, garnet) quickly dissolve into the melt and Al therefore behaves as a moderately incompatible element. During low-pressure fractionation of basalts, Al is removed by plagioclase precipitation. At higher pressure, plagioclase solubility in silicate melts increases, and this mineral does not precipitate until a late stage in the magmatic differentiation. Aluminum is therefore useful in assessing the depth of differentiation of basaltic series. In mid-ocean ridge and continental flood basalts, Al concentrations do not vary much with fractionation because they are buffered by plagioclase removal: these lavas are differentiated at low pressure in the plagioclase stability field. In contrast, Al concentrations increase steadily with fractionation of Hawaiian basalts: these rocks evolve at higher pressure in the absence of plagioclase. Typical concentrations of Al_2O_3 in basaltic and granitic melts average 15 wt %.

Aluminum solubility in hydrous fluids is low, except at very high temperature and high pH. This solubility is controlled by the stability of different complexes with the hydroxyl ion OH^- and the solubility of clay minerals in equilibrium with the solution, e.g. kaolinite. The minimum in solubility at pH $\approx$ 6 reflects the amphoteric character of this element. During weathering, Al solubility is controlled by kaolinite reactions of alkalinity production (6.15). The destruction of feldspars by CO_2-rich fresh water leaves Al in clays: this element is most efficiently transported by rivers to the sea in the suspended load. In seawater, Al solubility is controlled by the input of airborne clay particles transported from deserts, which gives Al distribution in the water column an unusual profile with concentration decreasing down the thermocline (Fig. 6.12).

10.3 Potassium

Most common form: K^+
Ionic radius: 1.51 Å (octahedral) and 1.64 Å (dodecahedral)
Stable isotopes: 39 (93.26%), 40 (0.011%), 41 (6.73%)
Atomic weight: 39.098
Long-lived isotope: 40 ($T^{1/2} = 1.25 \times 10^9$ a)
Condensation temperature: 1000 K
No significant complex in waters
Residence time in seawater: 12×10^6 years

Potassium is an alkaline element i.e. both volatile and lithophile. Its concentration in the Earth is therefore poorly constrained. As for other volatile elements, K is depleted in the Earth with respect to carbonaceous chondrites and enriched with respect to Mars and the Moon. The dual decay of the radioactive isotope ^{40}K into either ^{40}Ca by β^- emission or ^{40}Ar by electron capture makes this element, after U and Th, one of the three significant sources of heat in the Earth and accounts for about 20% of the radioactive heat production. As discussed in Chapter 8, it is not known whether K enters into the core composition.

The mineral phases that host K in the mantle are not well understood. Under upper mantle conditions, traces of high-K mineral phases, notably the Mg-rich mica phlogopite and occasionally alkali feldspar, may be present in those parts of the mantle that have been contaminated by subducted sediments or by fluids produced by the dehydration of the subducted oceanic crust. In the lower mantle, K may at least in part reside in the mineral hollandite, a high-pressure equivalent of alkali feldspar. The prime K repositories in igneous rocks are alkali feldspars, amphibole, and micas. In sedimentary rocks, most of the K inventory is sequestered in clay minerals, notably in smectites, illite, and in detrital feldspars.

The K^+ ion is very large. Although it may enter octahedral sites, it often fits 12-coordinated sites as in feldspars and micas. During melting in the mantle and basalt differentiation, K is strongly incompatible and follows other incompatible elements such as Th and U to the extent that the K/U ratio remains essentially constant in the mantle and the crust. Mid-ocean ridge basalts are far less concentrated ($\approx$0.1 wt%) in K than ocean island basalts (1–2 wt%). Granitic melts typically contain 2–3 wt% K_2O. Felsic igneous melts are normally saturated in feldspar and biotite: K is therefore not incompatible in the genesis of granitic rocks.

At low temperature and therefore during weathering, feldspars react with water to produce clay minerals. The K-rich clay mineral illite is left as a residue and its ubiquitous presence keeps the concentration of K in low-temperature hydrous fluids (rivers, seawater) at very low levels. K is therefore transported to the sea with the suspended load. In high-temperature hydrothermal fluids, K reacts with country rocks to form the K-feldspar crystals commonly observed in the metamorphic aureoles of granitic intrusions. Upon cooling, K is leached preferentially to Na, while the converse becomes true when the temperature of interaction between fluids and their ambient rocks falls below about 300 °C.

Solubility of KCl (sylvite, a major fertilizer) in water increases greatly with temperature. Through evaporation of brines in cool and cold ($<50°$) landlocked or lagoonar environments, sylvite becomes saturated before NaCl (halite). In seawater, K is not biolimited and its concentration is essentially constant in the water column.

10.4 Sodium

> Most common form: Na^+
> Ionic radius: 1.02 Å (octahedral)
> Stable isotope: 23 (100%)
> Atomic weight: 22.990
> Condensation temperature: 970 K
> No significant complex in waters
> Residence time in seawater: 83×10^6 years

Sodium is a volatile and lithophile alkaline element. As with K, Na terrestrial abundance is therefore not well known. Breaking down the inventory of Na among mantle minerals is very difficult. Most major mineral phases do not accept this element but, under upper mantle conditions, Na solubility in clinopyroxene increases substantially with pressure. In the crust, Na is essentially hosted in the albite component of plagioclase feldspar. In contrast with potassium, there is no major Na-rich clay mineral of major geological importance and most sedimentary Na resides in detrital feldspar. Evaporitic rock salt NaCl represents another significant surface repository of sodium.

In silicates, the Na^+ ion normally occupies octahedral sites. It is a moderately incompatible element during mantle melting and the early stages of basalt differentiation. Plagioclase saturation in basalts differentiated at low pressure such as mid-ocean ridge and continental flood basalts (see Section 10.2) and in more felsic melts turns Na into a moderately compatible element. Basalts typically contain 2–3 wt% Na_2O and granitic melts 3–4 wt % Na_2O. Although albite may be stable under near-surface conditions, it rarely reaches saturation in hydrous fluids at ambient temperature. Authigenic albite is known in sediments, but is not ubiquitous. Na is therefore preferentially leached by low-temperature hydrothermal fluids in contrast to K, which dominates in higher temperature fluids. Alteration of seafloor basalts by warm percolating fluids rich in sodium and silica leads to the replacement of calcic plagioclase (anorthite) by albite:

$$3CaAl_2Si_2O_8 + 6Na^+ + 12SiO_2 = 6NaAlSi_3O_8 + 3Ca^{2+}$$

Basalts altered on the seafloor are therefore notably enriched in Si and Na with respect to their unaltered precursor. There is a stark contrast between K and Na behavior under surface conditions imposed by the lack of a major stable low-temperature Na-rich phase. Sodium ions set free by the weathering of feldspars are transported by rivers into the sea in the dissolved load. Sodium is the second most abundant element in seawater and the most abundant cation. Its concentration is limited mostly by interaction of seawater with the seafloor through diagenetic and hydrothermal fluids. Its concentration is constant throughout the water column. In warm landlocked or lagoonar environments, NaCl saturation may be reached and evaporitic halite becomes a prominent sedimentary deposit.

10.5 Magnesium

Most common form: Mg^{2+}
Ionic radius: 0.72 Å (octahedral)
Stable isotopes: 24 (78.99%), 25 (10.00%), 26 (11.01%)
Atomic weight: 24.305
Condensation temperature: 1340 K
Complexes in water: hydroxides, carbonates, sulfates
Reactions limiting solubility in water:

$$CaMg(CO_3)_2 \Leftrightarrow Ca^{2+} + Mg^{2+} + 2CO_3^{2-} \; (-\log K = 1.70) \text{ (dolomite)}$$

Residence time in seawater: 13×10^6 years

Magnesium is a refractory and lithophile alkaline-earth element. It does not enter core composition in substantial concentration. After oxygen, magnesium is the second most abundant element in the mantle where it is hosted in most major minerals (olivine, pyroxenes, garnet, spinel, ringwoodite, etc.). Its concentration in the average crust is relatively low: as a consequence, amphiboles, micas, Mg-rich clays (smectite), and carbonates (dolomite) normally remain minor mineral phases. Mg-rich calcite is a major form of carbonate precipitated by marine organisms. After Na, Mg is the second most abundant cation of seawater in which it is partly complexed by carbonate and sulfate ions.

In silicates, Mg ions occupy octahedral sites. Many of its properties are better understood by observing that Fe and Mg have identical ionic charges and similar ionic radii. In sedimentary and igneous carbonates, Mg substitutes for Ca. During mantle melting and magma differentiation, Mg is strongly compatible. Primary basalts are rich in Mg and rapidly lose most of it by gravitational removal of olivine and clinopyroxene. The most common indices of magmatic differentiation involve Mg, either as the MgO content, the FeO/MgO ratio or the fraction known as the mg# and which is the atom ratio Mg/(Fe + Mg) $\approx 100/(1 + 0.56 \text{ FeO/MgO})$. Relatively undifferentiated basalts may contain 8–16 wt % MgO with FeO/MgO ratios close to unity and mg# in excess of 65. Typical granitic melts only contain 2–4 wt % MgO. Probably the most striking property of Mg in hydrous fluids is its strong shift in solubility between ambient temperature and about 70 °C: seawater, cold ground water, and runoff contain substantial amounts of Mg, while all hydrothermal and warm diagenetic fluids are Mg-free. A number of processes help explain such a shift: precipitation of acid Mg-sulfate in the black smoker feeding zones (see Chapter 7), precipitation of Mg-Ca carbonate in coastal environments (dolomitization), etc. The removal of seawater Mg in the ridge-crest hydrothermal systems approximately balances out riverine input of Mg to the ocean.

Magnesium is liberated from silicates by weathering (6.23) and transported to the sea in the dissolved load. Its abundance is not limited by a low-solubility mineral phase and is essentially constant down the water column. Partitioning of Mg into calcium carbonate seems to be dependent on the temperature conditions of precipitation: the Mg content of carbonates can therefore be used as a thermometer of ancient marine environments.

10.6 Calcium

Most common form: Ca^{2+}
Ionic radius: 1.00 Å (octahedral)
Stable isotopes: 40 (96.94%), 42 (0.65%), 43 (0.14%), 44 (2.09%), 46 (0.004%), 48 (0.19%)
Atomic weight: 40.078
Condensation temperature: 1518 K
Complexes in water: hydroxides, carbonates, sulfates
Reactions limiting solubility in water:

$$CaCO_3 \Leftrightarrow Ca^{2+} + CO_3^{2-} \ (-\log K = 8.22) \ (\text{aragonite})$$
$$CaCO_3 \Leftrightarrow Ca^{2+} + CO_3^{2-} \ (-\log K = 8.22) \ (\text{calcite})$$
$$CaSO_4.2H_2O \Leftrightarrow Ca^{2+} + SO_4^{2-} + 2H_2O \ (-\log K = 4.62) \ (\text{gypsum})$$

Residence time in seawater: 1.1×10^6 years

Calcium is a refractory and lithophile alkaline-earth element. The abundance of the isotope 40 produced by radioactive decay of ^{40}Ar is occasionally used as a chronometer. The relative abundances of non-radiogenic isotopes have also been used as a tracer of certain biological processes. Just like Mg, Ca does not enter core composition in substantial quantities. In the mantle, Ca is stored in clinopyroxene and in its high-pressure equivalent Ca-perovskite. In igneous rocks, as in the crust in general, calcic plagioclase (anorthite) and amphibole are major hosts for calcium. There is no major Ca-rich clay mineral. The major low-temperature Ca-rich phase is Ca carbonate in its two forms calcite and aragonite. Calcium phosphates are a ubiquitous form of Ca storage in igneous (fluorapatite) and sedimentary rocks (carbonate-apatite). Apatite is the essential ingredient of vertebrate hard parts (bones and teeth). Calcium sulfates (gypsum and anhydrite) are an essential component of evaporitic sequences.

In silicates, Ca ions occupy octahedral sites. During mantle melting, Ca is slightly incompatible. Its behavior during magmatic differentiation is controlled by the stability of plagioclase and clinopyroxene (see Section 10.2). High-pressure magmatic differentiation takes place in the presence of clinopyroxene but in the absence of plagioclase and therefore decreases the Ca/Al ratio in the residual melt. Low-pressure fractionation of basalts in the stability field of clinopyroxene and plagioclase leaves the Ca/Al ratio essentially constant. Although the Ca content of olivine is low (a fraction of a percent), it increases with decreasing temperature. Olivine phenocrysts in basalts may contain up to 0.5 wt% CaO, while mantle peridotites are normally far more depleted. Basaltic melts typically contain 10–12 wt% CaO and a granitic melt 2–4 wt% CaO.

Calcium is a mobile element during water–rock interaction at any temperature. Plagioclase, clinopyroxene, and amphibole are easily weathered by water equilibrated with atmospheric CO_2 and by seawater alteration of submarine basalts. Ca^{2+} goes into solution and is transported to the sea by runoff, while Ca-free mineral phases, silica and clay minerals, are left behind. Carbonate, sulfate, and phosphate complexes of Ca are strong. Calcium is the third most concentrated cation of seawater. The Ca concentration level in fresh water and seawater is controlled by the solubility of calcite and aragonite. Calcite is less soluble than aragonite. Surface oceanic water is saturated in calcite, while deep water is undersaturated. This results from the combined effect of temperature and pressure on the solubility product of calcium carbonate and the dissociation constants of carbonic acid but also of the ΣCO_2 content of local seawater. Removal of calcium carbonate from the ocean by biogenic carbonate precipitation is the most important control of seawater alkalinity and therefore is the crux of the response of the ocean–atmosphere system to CO_2 fluctuations due to changes in volcanic activity, biological productivity, and erosion patterns.

In brines from landlocked and lagoonar environments, precipitation of calcium sulfate may occur as either hydrated gypsum at temperatures below about 35 °C or anhydrous anhydrite at higher temperatures. Gypsum normally converts to anhydrite during burial.

10.7 Iron

Most common forms: Fe^0, Fe^{2+}, and Fe^{3+}
Ionic radius: 0.61 Å for Fe^{2+} and 0.55 Å for Fe^{3+} (octahedral)
Stable isotopes: 54 (5.90%), 56 (91.72%), 57 (2.10%), 58 (0.28%)
Atomic weight: 55.847
Condensation temperature: 1336 K

Complexes in water: hydroxides, chlorides
Reactions limiting solubility in water:

$$Fe(OH)_3 \Leftrightarrow Fe(OH)_2^+ + OH^- \ (-\log K = 16.5)$$
$$Fe(OH)_3 + H_2O \Leftrightarrow Fe(OH)_4^- + OH^- \ (-\log K = 4.4)$$

Residence time in seawater: 55 years

Iron is the most abundant element in the Earth. It is refractory and, by definition, siderophile. It is the most abundant element of the core, both in the solid inner core and the liquid outer core in which convection generates the terrestrial magnetic field. In contrast with the core where Fe occurs in its metallic and most reduced form Fe^0, iron in the mantle is essentially in its Fe^{2+} form. Ferrous iron (Fe^{2+}) substitutes for Mg^{2+} in most silicate mineral phases. Fe is, after Si and Mg, the third most abundant cation in the mantle. In the upper mantle, ferrous iron is found in olivine, pyroxene, garnet, and amphibole. In the deep mantle, it enters with Mg into the perovskite structure of ringwoodite but also into the oxide structure of magnesio-wüstite $(Fe, Mg) O$. In igneous rocks, as in the crust in general, it is hosted in amphibole and biotite but also, together with Fe^{3+}, Al^{3+}, Cr^{3+}, and Ti^{4+}, in oxide minerals (magnetite, ilmenite). Ferric iron easily substitutes into the tetrahedral site of alkali feldspars, which is why so many granites turn reddish upon incipient weathering. When exposed to the atmosphere or seawater at low temperature, Fe is normally oxidized to Fe^{3+}. It is found in different forms of iron hydroxide (such as goethite, hematite, and limonite) that dominate soils, sediments, as well as ferromanganese nodules and encrustations from the deep sea. Iron-rich clay minerals and carbonates are uncommon. Organic compounds contain important Fe-rich proteins that have different functions, notably oxygen transport in the cell (porphyrins). Iron concentration in seawater is very low, again because of the very low solubility of hydroxides.

The sites occupied by Fe^{2+} and Fe^{3+} are normally octahedral, but Fe^{3+} can be found in tetrahedral sites, especially in feldspars. During mantle melting, Fe^{2+} has a neutral behavior (neither compatible nor incompatible) owing to its lack of octahedral site preference energy (see Chapter 1). In contrast, Fe^{3+} is highly incompatible. Silicates are very poor electrical conductors so magmas cannot exchange significant amounts of electrons with country rocks. Upon removal of ferrous iron into cumulate minerals, the Fe^{3+}/Fe^{2+} ratio therefore increases in residual melts. The apparently more oxidizing conditions of differentiated rocks are therefore not the result of an externally imposed higher "fugacity" of oxygen, but simply reflect the increasing electron deficit in smaller and smaller quantities of melt. Most basalts would contain 10–12 wt% total iron as FeO and granites 3–4 wt%. Typically, about 15 wt% of the iron present in a primary basalt is in the form of Fe^{3+}. Even when small quantities of ilmenite and magnetite are present at the liquidus, iron concentration in melts does not change when the fractionating assemblage is dominated by olivine and pyroxene: this is the case for the high-pressure differentiation of ocean island basalts and for the wet differentiation of orogenic (calc-alkaline) magmas. When plagioclase is present, as in mid-ocean ridge and continental flood basalts (see Section 10.2), iron is significantly more incompatible. These contrasting trends are known in the petrological literature as the Fenner trend (constant Fe) and the Bowen trend (increasing Fe), respectively.

The behavior of iron during water–rock interaction must be understood with respect to the different properties of Fe^{2+} and Fe^{3+} in solution. Although Fe^{3+} is strongly complexed by Cl ions, its concentration is greatly limited by the solubility of ferric iron hydroxides. In contrast, Fe^{2+} is highly soluble. Reducing solutions can transport huge amounts of ferrous Fe and reprecipitate it when the conditions become more oxidizing. This is the case of the hydrothermal solutions of the submarine black smokers that precipitate enormous amounts of hydroxides and form the metalliferous sediments observed in the vicinity of the mid-ocean ridges. This is also the case of the Archean oceans that precipitated the banded iron formations (BIF) that form our current prime iron ore: the low pressure of oxygen in the ancient atmosphere permitted the long-distance transport of ferrous iron until subtle changes in the marine oxidizing conditions triggered the massive precipitation of iron hydroxide. The modern atmosphere being oxygen-rich, iron is kept in its ferric form. Weathering of the continental crust leaves residues of hydroxide minerals that are transported to the sea with the riverine suspended load. Some of these hydroxides are in the form of colloids of very small dimension (smaller than the 0.45 micrometer pore dimension of common filtering devices). The very large surface area and the charged surface of these colloids allows their mutual repulsion and therefore maintains the colloids in suspension. Adsorption of many highly charged elements on these colloids is very efficient. This is the case of many transition elements such as Cr^{3+}, but also of all the rare-earth elements (3+), uranium, thorium, etc., that get their ride to the sea on these colloids. When rivers meet seawater in estuaries, the dielectric properties of the mixture change allowing the particles to flocculate and precipitate with clay minerals as estuarine sediments, entraining in this process a very large fraction of the highly charged elements before they reach the open ocean.

Iron is mostly introduced into the ocean as airborne particles of goethite and limonite transported by winds from arid areas (Sahara, Gobi). Its concentration profile in the water column is unusual (and is only matched by that of cobalt). It displays concentrations decreasing from the surface to the bottom, which is a result of the progressive dissolution of the falling airborne particles. Iron may play an important role in the control of biological productivity in the ocean, and, via CO_2 consumption by phytoplankton, in the control of the greenhouse effect and climates. Some scientists believe that Fe, which is a trace nutrient (micronutrient) indispensable to biological activity, is normally in short supply in surface waters. In this scenario, the introduction of Fe-rich airborne particles would fertilize the sea more efficiently during dry periods and quickly draw down part of the atmospheric CO_2.

The behavior of iron at the interface between seawater and the oceanic crust also represents a set of major geochemical processes. Upon circulation of seawater through the fractured oceanic crust, basaltic Fe^{2+} is oxidized into Fe^{3+} at the expense of dissolved O_2, then through the re-duction of seawater SO_4^{2-} into sulfide S^{2-} (see Chapter 7). During diage-nesis, a different process takes place in the oxygen-starved layers below the bioturbation layer: biological reduction of SO_4^{2-} into sulfide S^{2-} changes the redox conditions. Ferric iron hydroxides (such as those com-posing ferromanganese nodules) are reduced to soluble Fe^{2+}, which then precipitates with S^{2-} to form the pyrite Fe_2S commonly found in reduced sediments.

10.8 Sulfur

Most common forms: S^0, S^{2-}, and SO_4^{2-}
Ionic radius: 0.31 Å (tetrahedral), 0.29 Å (octahedral), and 1.84 Å (S^{2-})
Stable isotopes: 32 (95.02%), 33 (0.75%), 34 (4.21%), 36 (0.02%)
Atomic weight: 32.07
Condensation temperature: 648 K
Dissociation of H_2S in water:

$$H_2S \Leftrightarrow HS^- + H^+ \ (-\log K = 7.02)$$
$$HS^- \Leftrightarrow S^{2-} + H^+ \ (-\log K = 13.9)$$

Reactions limiting solubility in water:

$$CaSO_4.2H_2O \Leftrightarrow Ca^{2+} + SO_4^{2-} + 2H_2O \ (-\log K = 4.62) \ (\text{gypsum})$$
$$FeS \Leftrightarrow Fe^{2+} + S^{2-} \ (-\log K = 18.1) \ (\text{pyrrhotite})$$

Residence time in seawater: not known but oceans are well mixed for sulfate

Sulfur is strongly chalcophile and volatile. It has been repeatedly sug-gested that very large quantities of this element are dissolved in the core and contribute to the relatively low seismic velocities of the core with respect to those of pure iron. Sulfur does not readily dissolve in silicates. Terrestrial sulfur is therefore stored in sulfides. At high temperatures, solid solutions of Ni and Fe dominate (monosulfide solid solution, MSS). At ambient temperature, sulfur enters a variety of sulfides. The major repository of sulfur in sediments is sulfides, notably pyrite. Because sulfates, the oxidized form of sulfur, are relatively soluble, these minerals play a minor role in the making of continental crust, with the exception of gypsum and anhydrite in evaporites, and barite in hydrothermal veins. In seawater, river, and rain-water, in which substantial amounts of dissolved oxygen are present, the stable form of sulfur is the oxidized form, sulphate SO_4^{2-}, which is the third most abundant ion of seawater.

In magmas, sulfur is present as sulfide and also as sulfate in more oxidized granites. During mantle melting and magma differentiation, sulfur has a compatible behavior controlled by the exsolution and gravitational segregation of sulfide blebs (the melting point of sulfides is lower than average magma temperature) that are occasionally observed as mineral inclusions. Sulfides are the most abundant minerals of inclusions in diamond. The high-temperature form of sulfides precipitated out of mafic magmas are characteristically rich in nickel (pentlandite). The high-temperature forms of igneous sulfides are very unstable with respect to low-temperature alteration.

Weathering oxidizes sulfides into soluble sulfate and all the sulfur from the crust is transported to the sea dissolved in the runoff. In landlocked and coastal environments, evaporation concentrates natural waters eventually to the point of gypsum and anhydrite saturation: these minerals are the most abundant form of evaporites. In seawater, $BaSO_4$ saturation is often reached, but most of it is re-dissolved shortly after deposition. Microbial activity during early diagenesis turns marine SO_4^{2-} from interstitial fluids into sedimentary pyrite. Marine SO_4^{2-} introduced into submarine hydrothermal systems (black smokers) is first precipitated as anhydrite whose solubility decreases significantly with temperature, reduced by ferrous iron from ambient basalt, and mixed with sulfide leached from the basalts. The reduced forms H_2S and HS^- are present in hydrothermal fluids, largely leached from the basalts, in proportions that vary largely with the pH. Upon cooling, hydrothermal sulfur commonly precipitates as sulfides of Cu and Fe (chalcopyrite, pyrrhotite) at temperatures in excess of $300\,°C$, and of Zn (sphalerite) at lower temperatures. Sulfur is abundant in volcanic fumaroles and is released as SO_2 in large quantities into the atmosphere by volcanic eruptions. An important aspect of sulfur atmospheric chemistry is the production of gaseous dimethyl-sulfide (DMS) generated in vast quantities by phytoplankton.

10.9 Phosphorus

Most common form: PO_4^{3-}

Ionic radius: 0.17 Å (tetrahedral)

Stable isotopes: 31 (100%)

Atomic weight: 30.974

Condensation temperature: 1151 K

Dissociation of H_3PO_4 in water:

$$H_3PO_4 \Leftrightarrow H_2PO_4^- + H^+ (-\log K = 2.15)$$
$$H_2PO_4^- \Leftrightarrow HPO_4^{2-} + H^+ (-\log K = 7.20)$$
$$HPO_4^{2-} \Leftrightarrow PO_4^{3-} + H^+ (-\log K = 12.35)$$

Reaction limiting solubility in water:

$$Ca_5(PO_4)_3OH \Leftrightarrow 5\ Ca^{2+} + 3PO_4^{3-}$$
$$+ OH^- (-\log K = 55.6)\,(\text{hydroxylapatite})$$

Residence time in seawater: 70 000 years

Phosphorus is a lithophile and moderately siderophile element. Substantial amounts of this element are probably dissolved in the liquid core. It is almost exclusively hosted in Ca phosphate (apatite) (see Section 10.6). Apatite may be of igneous origin. Although apatite is certainly present in the upper mantle, P repository in the deep mantle is not well understood. Biogenic (fish teeth and bones) and diagenetic apatites are the essential repositories of sedimentary phosphorus. They occasionally form huge deposits, as in West Africa, that are actively mined to provide agricultural fertilizer. Some of these deposits, found in particular in the Late Precambrian of China, are chemical precipitates and seem to be associated with episodes of global glaciation. In low-temperature waters, phosphates form numerous complexes and, as indicated by the dissociation reactions above, speciation is pH-dependent. Phosphorus concentration in seawater and river water is limited by the very low solubility of apatite. Phosphate radicals often attach themselves to the surface of iron oxi-hydroxide colloids when they precipitate in estuaries.

Phosphorus occurs principally in the center of oxygen tetrahedra. In seawater, P is one of the essential nutrients: in the Krebs cycle, adenosine triphosphate (ATP) is the major energy repository for cells, while ribose phosphates form the building blocks of nucleic acids DNA and RNA. Phosphate is brought to seawater by rivers and is removed, though only after many cycles across the thermocline, with the hard parts of biogenic debris. It is severely depleted in surface water but is regenerated during the dissolution of falling debris in deep water. It is depleted in young North Atlantic Deep Water with respect to the older Antarctic Bottom

Water. Diagenetic dissolution, transport, and re-precipitation of apatite is common in sediments.

10.10 Carbon

Carbon is not an extremely abundant component of the Earth. It is both siderophile (in its reduced form) and atmophile (in its oxidized form). Substantial amounts of this element may be dissolved in the core. In the mantle, carbon occurs as graphite and, at depth in excess of about 120 km, as diamond when the conditions are reducing. In oxidizing conditions, carbon occurs as carbon dioxide, which, at depths of about 70–100 km, reacts with mantle silicates to form carbonates, e.g:

$$Mg_2SiO_4 + CO_2 \Leftrightarrow \quad MgCO_3 \quad + \quad MgSiO_3$$
$$\text{(olivine)} \qquad\qquad \text{(magnesite)} \quad \text{(pyroxene)} \tag{10.1}$$

Carbon dioxide solubility in magmas rapidly changes with pressure and therefore depth. CO_2 outgassing from mantle-derived magmas starts at a depth of approximately 60 km and quickly strips the magma of many volatile species, such as rare gases, well before eruption. CO_2 is found as fluid inclusions in olivine phenocrysts (the prime target for He isotope measurements) and makes up a very important component of the gas phases in MORB and OIB.

   Oxidized carbon in the crust occurs as sedimentary carbonates, mostly calcite which is two–three times more abundant than dolomite. The reduced forms are countless, from crystalline graphite to amorphous organic varieties (coal, oil, kerogen, methane). CO_2 is an important component of metamorphic gases and, in particular, is the dominant species in fluids from the granulite facies. In ground-, river-, and seawater, carbon occurs as carbonate oxi-anions CO_3^{2-} and HCO_3^-, but traces of soluble organic components (humic and fulvic acids) are ubiquitous. Carbon dioxide makes up about 350 ppm per volume of the atmospheric gases.

   The geochemical cycle of carbon is of particular significance because it is the most essential component of life. The extremely diverse polymerization modes of carbon compounds and their easy binding to a number of other elements and molecules (nitrogen, phosphate, iron, magnesium, and scores of other metals) are unique in nature. A major source of carbon dioxide is volcanic outgassing from the mantle. Carbonate groups are essentially indestructible except by biological activity or in extremely reducing environments. The prime sites for production of reduced carbon from atmospheric CO_2 and oceanic carbonates are the ocean surface, continental shelves, and continental biosphere. This production, fuelled by photosynthetic processes, is called primary productivity. Igneous and biogenic reduced carbon is easily oxidized by atmospheric oxygen during weathering. The resulting carbon dioxide is distributed almost equally between atmospheric CO_2 and oceanic carbonates. Burial and subduction of sediments rich in organic carbon (reverse weathering) leaves unbalanced oxygen that accounts for the high proportion of this gas in the atmosphere.

The short-term components (ten–100 000 years) of the carbon cycle are largely driven by fluctuations in biological activity and are affected by human activities such as coal and oil burning, land use, and deforestation. This aspect will be left to more specialized monographs and textbooks.

Reference

Wasson, J. T. (1985) *Meteorites: Their Record of Early Solar System History*. Berlin, Springer.

Appendix A
Composition of the major geological units

Table A.1: *The table below recaps the composition of the Earth's main reservoirs and of the CI carbonaceous chondrites, which serve as a reference for the composition of the Earth. The composition of the silicate Earth, representing the Bulk Silicate Earth less the core and fluid envelopes, is discussed in the text. The concentration of very many elements (nutrients) in the ocean is also highly variable and the values are indicative only. ppm = parts per million, ppb = parts per billion. BSE = Bulk Silicate Earth. A large amount of data can be found on the Geochemical Earth Reference Model (GERM) internet site: http://www.earthref.org.*

Element	N	M at	Chondr. CI	Earth	Core	BSE	Cont. crust	Upper mantle	Unit	Ocean $(\mu g\, l^{-1})$	Rivers $(\mu g\, l^{-1})$
H	1	1.01	20000	67.5	600	100		2.5	ppm	1.11×10^8	
He	2	4.00								7.20×10^{-3}	
Li	3	6.94	1.5	1.08	0	1.6	11	1.568	ppm	1.78×10^2	3.0×10^0
Be	4	9.01	0.025	0.046	0	0.068		0.0442	ppm	2.25×10^{-4}	1.0×10^{-2}
B	5	10.81	0.900	0.203	0	0.3		0.075	ppm	4.39×10^3	
C	6	12.01	35000	309	2000	120		18	ppm	2.64×10^4	
N	7	14.01	3180	27	75	2		0.05	ppm	8.54×10^3	
O	8	16.00	46	30.5	0	44	45	44	%	2.40×10^3	
F	9	19.00	60	10.1	0	15		9.75	ppm	1.27×10^3	1.0×10^0
Ne	10	20.18								3.96×10^{-4}	
Na	11	22.99	0.510	0.180	0	0.267	2.37	0.214	%	1.08×10^7	6.0×10^3
Mg	12	24.31	9.65	15.39	0	23	2.65	22.8	%	1.26×10^6	4.1×10^3
Al	13	26.98	0.86	1.59	0	2.35	8.36	2.23	%	1.62×10^{-1}	5.0×10^1
Si	14	28.09	10.7	17.10	6	21	27.61	21	%	2.81×10^3	6.5×10^3
P	15	30.97	1080	1101	3500	90	873	54	ppm	6.00×10^1	2.0×10^1
S	16	32.07	54000	6344	19000	250		238	ppm	8.63×10^5	3.7×10^3
Cl	17	35.45	680	11.5	200	17		2.55	ppm	1.89×10^7	7.8×10^3

Table A.1 (*cont.*):

Element	N	M at	Chondr. CI	Earth	Core	BSE	Cont. crust	Upper mantle	Unit	Ocean (μg l^{-1})	Rivers (μg l^{-1})
Ar	18	39.95								5.99×10^2	
K	19	39.10	550	162	0	240	15772	24	ppm	3.89×10^5	3×10^3
Ca	20	40.08	0.925	1.71	0	2.53	4.57	2.40	%	4.14×10^5	2×10^4
Sc	21	44.96	5.92	10.93	0	16.2	22	15.39	ppm	6.00×10^{-4}	4×10^{-3}
Ti	22	47.88	440	813	0	1205	4197	928	ppm	4.79×10^{-3}	3×10^0
V	23	50.94	56	94.4	150	82	131	82	ppm	1.78×10^0	9×10^{-1}
Cr	24	52.00	2650	3720	9000	2625	119	2625	ppm	2.08×10^{-1}	1×10^0
Mn	25	54.94	1920	1680	3000	1045	852	1045	ppm	1.92×10^{-1}	7×10^0
Fe	26	55.85	18.1	30.3	85	6.26	5.13	6.26	%	5.59×10^{-2}	4×10^1
Co	27	58.93	500	838	2000	105	25	105	ppm	2.00×10^{-3}	1×10^{-1}
Ni	28	58.69	10500	17600	52000	1960	51	1960	ppm	4.80×10^{-1}	3×10^{-1}
Cu	29	63.55	120	60.9	125	30	24	29.1	ppm	2.00×10^{-1}	7×10^0
Zn	30	65.39	310	37.1	0	55	73	55	ppm	3.79×10^{-1}	2×10^1
Ga	31	69.72	9.20	2.70	0	4.000	16	3.8	ppm	6.97×10^{-4}	9×10^{-2}
Ge	32	72.61	31	7.24	20	1.100		1.1	ppm	5.08×10^{-2}	5×10^{-3}
As	33	74.92	1.850	1.669	5	0.050		0.01	ppm	1.72×10^0	2×10^0
Se	34	78.96	21	2.65	8	0.075		0.07125	ppm	1.18×10^{-1}	6×10^{-2}
Br	35	79.90	3.57	0.034	0.7	0.050		0.005	ppm	6.70×10^4	2×10^1
Kr	36	83.80								2.93×10^{-1}	
Rb	37	85.47	2.30	0.405	0	0.600	58	0.0408	ppm	1.24×10^2	1×10^0
Sr	38	87.62	7.25	13.4	0	19.865	325	12.935	ppm	8.00×10^3	7×10^1
Y	39	88.91	1.570	2.90	0	4.302	20	3.655	ppm	8.66×10^{-3}	4×10^{-2}
Zr	40	91.22	3.82	7.07	0	10.467	123	6.195	ppm	1.20×10^{-2}	
Nb	41	92.91	0.240	0.444	0	0.658	12	0.11186	ppm	4.65×10^{-3}	
Mo	42	95.94	900	1660	5000	50		30	ppb	1.10×10^1	6×10^{-1}
Ru	44	101.07	710	1310	4000	4.97		4.97	ppb		
Rh	45	102.91	130	242	740	0.910		0.91	ppb		
Pd	46	106.42	550	1015	3100	3.850		3.85	ppb	4.26×10^{-5}	
Ag	47	107.87	200	54.4	150	8		7.84	ppb	2.37×10^{-2}	3×10^{-1}
Cd	48	112.41	710	76.0	150	40		39.2	ppb	6.74×10^{-2}	1×10^{-2}
In	49	114.82	80	6.8	0	10		8.5	ppb	1.15×10^{-4}	
Sn	50	118.71	1650	251.1	500	130		97.5	ppb	4.75×10^{-4}	4×10^{-2}
Sb	51	121.76	140	46.2	130	5.5		1.1	ppb	1.22×10^{-1}	7×10^{-2}
Te	52	127.60	2330	285.7	850	12		11.76	ppb	1.02×10^{-4}	
I	53	126.90	450	6.8	0.13	10		1	ppm	5.71×10^1	7×10^0
Xe	54	131.29								6.56×10^{-2}	
Cs	55	132.91	0.190	0.014	0.065	0.021	2.6	0.504	ppm	3.00×10^{-1}	2×10^{-2}
Ba	56	137.33	2.410	4.455	0	6.600	390	0.449	ppb	1.17×10^1	2×10^1
La	57	138.91	0.237	0.437	0	0.648	18	0.080	ppm	3.88×10^{-3}	5×10^{-2}
Ce	58	140.12	0.613	1.131	0	1.675	42	0.538	ppm	7.01×10^{-4}	8×10^{-2}
Pr	59	140.91	0.093	0.171	0	0.254	5	0.114	ppm	7.72×10^{-4}	7×10^{-3}
Nd	60	144.24	0.457	0.844	0	1.250	20	0.738	ppm	2.60×10^{-3}	2×10^{-3}

Table A.1 (*cont.*):

Element	N	M at	Chondr. CI	Earth	Core	BSE	Cont. crust	Upper mantle	Unit	Ocean (μg l^{-1})	Rivers (μg l^{-1})
Sm	62	150.36	0.148	0.274	0	0.406	3.9	0.305	ppm	5.10×10^{-4}	8×10^{-3}
Eu	63	151.97	0.056	0.104	0	0.154	1.2	0.119		1.34×10^{-4}	2×10^{-3}
Gd	64	157.25	0.199	0.367	0	0.544	3.6	0.430	ppm	7.94×10^{-4}	9×10^{-3}
Tb	65	158.93	0.036	0.067	0	0.099	0.056	0.080	ppm	1.81×10^{-4}	1×10^{-3}
Dy	66	162.50	0.246	0.455	0	0.674	3.5	0.559	ppm	9.47×10^{-4}	7×10^{-3}
Ho	67	164.93	0.055	0.101	0	0.149	0.76	0.127	ppm	2.79×10^{-4}	1×10^{-3}
Er	68	167.26	0.160	0.296	0	0.438	2.2	0.381	ppm	9.17×10^{-4}	4×10^{-3}
Tm	69	168.93	0.025	0.046	0	0.068	0.32	0.060	ppm	1.63×10^{-4}	6×10^{-4}
Yb	70	173.04	0.161	0.297	0	0.441	2	0.392	ppm	9.81×10^{-4}	4×10^{-3}
Lu	71	174.97	0.025	0.046	0	0.068	0.33	0.061	ppm	1.80×10^{-4}	6×10^{-4}
Hf	72	178.49	0.103	0.191	0	0.283	3.7	0.167	ppm	1.61×10^{-4}	
Ta	73	180.95	0.014	0.025	0	0.037	1.1	0.006	ppm	2.53×10^{-3}	
W	74	183.84	0.093	0.173	0.47	0.029		0.002	ppm	1.10×10^{-1}	3×10^{-2}
Re	75	186.21	40	75.3	0.23	0.280		0.270	ppm	7.20×10^{-3}	
Os	76	190.23	490	900	2750	3.430		3.430	ppm	1.70×10^{-6}	
Ir	77	192.22	455	835	2550	3.185		3.185	ppb	1.15×10^{-6}	
Pt	78	195.08	1010	1866	5700	7.070		7.070	ppb	1.17×10^{-4}	
Au	79	196.97	140	164	500	0.980		0.960	ppb	9.85×10^{-6}	2×10^{-3}
Hg	80	200.59	300	23.1	50	10		9.800	ppb	1.40×10^{-3}	7×10^{-2}
Tl	81	204.38	140	12.2	30	3.500		0.350	ppb	1.23×10^{-2}	
Pb	82	207.20	2.470	0.232	0.4	0.150	12.6	0.018	ppb	5.00×10^{-4}	1×10^{0}
Bi	83	208.98	110	9.853	25	2.500		0.500	ppb	1.04×10^{-5}	
Th	90	232.04	0.029	0.054	0	0.079	5.6	0.006	ppm	5.00×10^{-5}	
U	92	238.03	0.0074	0.014	0	0.020	1.42	0.002	ppm	3.20×10^{0}	4×10^{-2}

Data sources: Solid Earth: McDonough and Sun (1995); ocean: GERM website; rivers: mostly Martin and Meybeck (1979).

References

Martin, J.-M. and Meybeck, M. (1979) Elemental mass-balance of material carried by major world rivers. *Mar. Chem.*, **7**, pp. 173–206.

McDonough, W. F. and Sun, S.-S. (1995) The composition of the Earth. *Chem. Geol.*, **120**, pp. 223–253.

Appendix B
The mixing equation for ratios

Let us write (2.6) describing the conservation of elements (or isotopes) A and B in the mixture of components $j = 1, 2, \ldots$:

$$C_0^A = \sum_j f_j C_j^A \tag{B.1}$$

$$C_0^B = \sum_j f_j C_j^B \tag{B.2}$$

and then divide one by the other:

$$\left(\frac{C^A}{C^B} \right)_0 = \frac{f_1 C_1^A + f_2 C_2^A + \cdots}{f_1 C_1^B + f_2 C_2^B + \cdots}$$

$$= \frac{f_1 C_1^B}{f_1 C_1^B + f_2 C_2^B + \cdots} \left(\frac{C^A}{C^B} \right)_1$$

$$+ \frac{f_2 C_2^B}{f_1 C_1^B + f_2 C_2^B + \cdots} \left(\frac{C^A}{C^B} \right)_2 + \cdots \tag{B.3}$$

In the fractions of the right-hand side, we recognize proportions φ_1^B, $\varphi_2^B, \ldots$, of the denominator element B provided by components $1, 2, \ldots$, proportions that are written, for component 1, say:

$$\varphi_1^B = \frac{f_1 C_1^B}{f_1 C_1^B + f_2 C_2^B + \cdots} \tag{B.4}$$

Summing the previous equation for all components, it can be verified that the sum of the various φ_j^B is equal to unity. We therefore arrive back at (2.9):

$$\left(\frac{C^A}{C^B} \right)_0 = \sum_j \varphi_j^B \left(\frac{C^A}{C^B} \right)_j \tag{B.5}$$

For a binary mixture, allowing for closure, this equation becomes:

$$\left(\frac{C^A}{C^B}\right)_0 = \left(\frac{C^A}{C^B}\right)_1 + \varphi_2^B\left[\left(\frac{C^A}{C^B}\right)_2 - \left(\frac{C^A}{C^B}\right)_1\right] \qquad (B.6)$$

If the element in the denominator is the same for two ratios, e.g. for A/B and X/B, or if the two denominators are proportional, φ_2^B can be eliminated between the two corresponding equations. We thus obtain a linear relation between the two ratios for the mixture. For two independent denominators, say for A/B and X/Y, the relation can be shown to be hyperbolic.

Appendix C
A refresher on thermodynamics

The first principle postulates the existence of a conservative quantity, energy. Energy exists in several forms: chemical bonds, thermal agitation of ions (the sum of these two giving the internal energy U), potential energy, kinetic energy, etc. Energy can be transferred in several ways, by heat transfer δQ (propagation of agitation) or mechanical work δW (exchange of a quantity of motion with a pressure agent, such as the atmosphere or the weight of a column of rock). The first principle can be written in simplified form as:

$$\mathrm{d}U = \delta Q + \delta W = \delta Q - P\mathrm{d}V \tag{C.1}$$

where P is pressure and V is the volume of the system. Heat Q and work W are not themselves conservative properties.

The second principle of thermodynamics states that an isolated system drifts spontaneously toward the most probable state, i.e. a state where a maximum number of equivalent microscopic configurations are available to it. The measure of the number of equivalent configurations accessible to a system is its entropy S. The entropy of an isolated system can therefore only increase. However, possible arrangements of a system can be added or removed by adding or subtracting energy, as in a game of checkers where the potential for varied situations depends on the number of pieces that can be arranged on the board. For a pure heat exchange at constant volume, we write:

$$\mathrm{d}S = \frac{\delta Q}{T} + \sigma\,\mathrm{d}t \tag{C.2}$$

where the variable T, defined by this equation, refers to the absolute temperature (in kelvins) of the system, t to time, and σ to the entropy

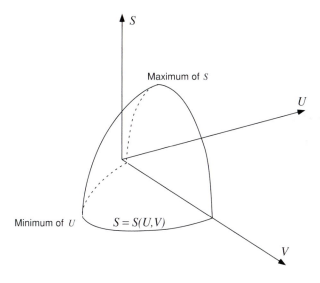

Figure C.1: The complementarity of the maximum entropy and minimum energy conditions are expressed by the single equation $S = f(U, V)$ representing the equation of a surface in (U, S, V) space. At constant energy U (i.e. for an isolated system), the volume V of the system evolves spontaneously so that its entropy is at a maximum. At constant entropy (no heat exchange with the outside), the same system evolves toward a minimum energy state.

production by dissipative processes, such as shear heating and chemical diffusion. According to the second principle, σ must be non-negative. For an infinitesimal reversible transformation, σ is zero. For a system whose state is controlled by prescribing its entropy S and its volume V, the variation in energy is described by the variation in internal energy U such that:

$$dU \leq TdS - PdV \tag{C.3}$$

which relation is reduced to an equality in the case of a reversible transformation. In (U, S, V) space, the entropy of a system whose energy remains constant evolves spontaneously toward a maximum (Fig. C.1). The system's geometry shows that an adiabatic system ($\delta Q = 0$) evolves spontaneously toward a minimum internal energy U.

For a system whose control variables are entropy and pressure, we use enthalpy $H = U + PV$. For a system for which the temperature T and pressure P are now prescribed, there is nothing to preclude an energy exchange with the outside, whether thermal or mechanical. The intrinsic energy of such a system is measured by another conservative magnitude, free enthalpy or Gibbs' free energy G, so that:

$$G = U + PV - TS \tag{C.4}$$

Other forms of energy are used when control variables other than T and P are preferred. Differentiating G and allowing for the definition of U gives:

$$dG = -SdT + VdP \tag{C.5}$$

An apparently exotic property of G can be readily derived from the previous equation and proves very useful for the study of chemical equilibria:

$$\left[\frac{\partial\,(G/T)}{\partial\,(1/T)}\right]_P = H \qquad\qquad (C.6)$$

For a perfect gas, the equation of state reads:

$$PV = nRT \qquad\qquad (C.7)$$

where n is the number of moles of gas in the enclosure and R is the gas constant. The minimum energy (maximum entropy) corresponds to $dG = 0$. At constant temperature and constant mole number, we can write:

$$dG = V dP = nRT\frac{dP}{P} \qquad\qquad (C.8)$$

giving upon integration:

$$G = nRT \ln P + g(T, n) \qquad\qquad (C.9)$$

where $g(T, n)$ is the Gibbs' free energy of n moles of gas at unit pressure. Let us now consider a system made up of several components (e.g. Na^+, Cl^-, and H_2O for a salt solution). The share of the total free enthalpy G of the system that can be assigned to each component i having n_i moles in the system is μ_i, which is obtained by writing:

$$G = \sum_i \mu_i n_i \qquad\qquad (C.10)$$

where the sum relates to all the components of the system. μ_i is known as the chemical potential of component i in the system. It is found by taking the derivative of this relation with respect to each variable that μ_i is simply the derivative of G relative to n_i, when T, P, and the number of moles of components other than i are kept constant. This yields the general expression of G for a system whose temperature, pressure, and composition are specified:

$$dG = -SdT + V dP + \sum_i \mu_i n_i \qquad\qquad (C.11)$$

If we consider a mixture of ideal gases, i.e. gas molecules of species other than i which do not interact with each other, the equation of state can be written as:

$$PV = \sum_i (n_i)RT \qquad\qquad (C.12)$$

where n is replaced by the sum of the numbers of moles of each species. The partial pressure P_i of gas i is defined as $(n_i/n)\,P$, and the Gibbs' free energy equation becomes:

$$G = \sum_i n_i \left[RT \ln P_i + \mu_i^0(T) \right] \tag{C.13}$$

where $\mu_i^0(T)$ is the chemical potential in the standard state $P_i = 1$ atm. By reference to (C.8), the chemical potential μ_i of gas i in the mixture of gases is defined as:

$$\mu_i = \mu_i^0(T) + RT \ln P_i \tag{C.14}$$

Let us now consider the following reaction in the gaseous state:

$$CH_4 + 2O_2 \Leftrightarrow CO_2 + 2H_2O \tag{C.15}$$

As the reaction progresses, the compounds are created and destroyed in proportions dictated by the stoichiometric coefficients:

$$\frac{dn_{CH_4}}{\nu_{CH_4}} = \frac{dn_{O_2}}{\nu_{O_2}} = \frac{dn_{CO_2}}{\nu_{CO_2}} = \frac{dn_{H_2O}}{\nu_{H_2O}} = d\xi \tag{C.16}$$

where $\nu_{CH_4} = -1$, $\nu_{O_2} = -2$, $\nu_{CO_2} = +1$, and $\nu_{H_2O} = +2$. The parameter ξ measures the progress of the reaction and varies between 0 and 1. The reaction will be at equilibrium when the free energy of the reaction products (right-hand side) is equal to that of the reactants (left-hand side), i.e. when the transfer of matter from one side to the other occurs with no change in the energy of the system. This condition is written by differentiating the expression for Gibbs' free energy as:

$$\mu_{CH_4} dn_{CH_4} + \mu_{O_2} dn_{O_2} + \mu_{CO_2} dn_{CO_2} + \mu_{H_2O} dn_{H_2O} = 0 \tag{C.17}$$

or alternatively:

$$\left(\mu_{CH_4} \nu_{CH_4} + \mu_{O_2} \nu_{O_2} + \mu_{CO_2} \nu_{CO_2} + \mu_{H_2O} \nu_{H_2O} \right) d\xi = \Delta G d\xi = 0 \tag{C.18}$$

which condition can only generally be observed when the content of the parentheses of the left-hand side cancels out ($\Delta G = \sum \mu_i \nu_i = 0$). By replacing the chemical potentials by their expression, we obtain:

$$RT \left[(-1) \ln P_{CH_4} + (-2) \ln P_{O_2} + (+1) \ln P_{CO_2} + (+2) \ln P_{H_2O} \right] =$$
$$- \left[(-1)\mu_{CH_4}^0 + (-2)\mu_{O_2}^0 + (+1)\mu_{CO_2}^0 + (+2)\mu_{H_2O}^0 \right] \tag{C.19}$$

which relation can be compacted using the properties of the logarithms to the form of the "mass action law:"

$$\ln \frac{P_{CO_2} P_{H_2O}^2}{P_{CH_4} P_{O_2}^2} = \ln K(T, P) = -\frac{\Delta G_0(T)}{RT} \tag{C.20}$$

In this equation, $K(T, P)$ is the equilibrium constant of the reaction and $\Delta G_0(T)$ the variation in Gibbs' free energy when all the components are in the standard state. Two essential equations accompany the mass action law and control the variation of the constant K with temperature and pressure. They are a consequence of the equations demonstrated above:

$$\left[\frac{\partial \ln K}{\partial (1/T)}\right]_P = -\frac{\Delta H_0(T, P)}{R} \tag{C.21}$$

$$\left[\frac{\partial \ln K}{\partial P}\right]_T = -\frac{\Delta V_0(T, P)}{RT} \tag{C.22}$$

This formalism established for ideal gases can be generalized to real gases by defining a parameter that satisfies the same equations as partial pressure; this is the gas fugacity.

It can also be transposed to liquid and solid solutions by replacing partial pressures by molar fractions $x_i = n_i/n$. For example, the substitution of rubidium (Rb) for potassium (K) between feldspar and mica can be described by the reaction:

$$Rb^+_{feld} + K^+_{mica} \Leftrightarrow Rb^+_{mica} + K^+_{feld} \tag{C.23}$$

The chemical potential of rubidium in feldspar is written:

$$\mu^{feld}_{Rb} = \mu^{feld,0}_{Rb}(T, P) + RT \ln x^{feld}_{Rb} \tag{C.24}$$

with similar expressions for other potentials. It will be noticed that the reference chemical potential $\mu^{feld,0}_{Rb} = 1$ is now defined for a unit molar concentration ($x^{feld}_{Rb} = 1$) and that it is dependent on temperature and pressure. The mass action law of equilibrium is written:

$$\ln \frac{x^{mica}_{Rb} x^{feld}_K}{x^{feld}_{Rb} x^{mica}_K} = \ln \frac{(Rb/K)_{mica}}{(Rb/K)_{feld}} = \ln K(T, P) = -\frac{\Delta G_0(T, P)}{RT} \tag{C.25}$$

In many cases like this, the trace element (Rb, a few tens of ppm) displaces only a tiny fraction of the major element (K, 10% or more). Molar fractions of potassium may be considered constant and we can write:

$$\ln \frac{x^{mica}_{Rb}}{x^{feld}_{Rb}} = \ln K'(T, P) \tag{C.26}$$

where K' is the partition coefficient of rubidium between mica and feldspar. In practice, it is equivalent to an equilibrium coefficient of the mass action law, particularly with respect to the equation controlling its dependence on temperature and pressure.

Finally, let us consider a heterogeneous equilibrium, i.e. which involves solid, liquid, and gaseous phases at the same time. For example, let us examine the decarbonation of a siliceous limestone:

$$\begin{array}{cccc} CaCO_3 + & SiO_2 & \Leftrightarrow & CaSiO_3 & + CO_2 \\ (calcite) & (quartz) & & (wollastonite) \end{array} \tag{C.27}$$

to produce wollastonite $CaSiO_3$, a silicate similar to pyroxene, and carbon dioxide. The equality of the chemical potentials can be written:

$$\ln \frac{x_{CaSiO_3} P_{CO_2}}{x_{CaCO_3} x_{SiO_2}} = -\frac{\Delta G_0(T, P)}{RT} \tag{C.28}$$

Assuming the minerals to be relatively pure, the molar fractions, such as x_{CaCO_3}, are all very close to unity, and we obtain the equation:

$$\ln P_{CO_2} = -\frac{\Delta G_0(T, P)}{RT} \tag{C.29}$$

We say that the calcite–carbonate–silica assemblage buffers the CO_2 pressure. For constant confining pressure P (pressure imposed by the rock column), (C.6) and (C.29) can be transcribed as:

$$\left[\frac{\partial \ln P_{CO_2}}{\partial(1/T)} \right] = -\frac{\Delta H_0(T, P)}{R} \tag{C.30}$$

where ΔH_0 is the latent heat (enthalpy) of reaction of the equilibrium in question. If the latent heat varies little within the temperature range considered, the equation can be integrated as:

$$\ln P_{CO_2} = -\frac{\Delta H_0(T, P)}{RT} + \text{constant} \tag{C.31}$$

Similar equations can be written for the water vapor pressure (dehydration reactions), oxygen pressure (oxidation–reduction reactions), and for many other species. Similar equations are also valid for solubility of a pure solid phase in a solution:

$$\left[\frac{\partial \ln x_i}{\partial(1/T)} \right] = -\frac{\Delta H_i(T, P)}{R} \tag{C.32}$$

where $\Delta H_i(T, P)$ is the heat absorbed by dissolution of a mole of component i in the solution (melting-point depression). It should be remembered that the non-ideal character of mixtures of gases and of solutions has been ignored all along.

We have seen that when an equilibrium is established between several components that react together, the change in Gibbs' free energy ΔG of the reaction is zero. By definition, this is equal to:

$$\Delta G = \Delta H - T \Delta S \tag{C.33}$$

where $\Delta H d\xi$ is the change in enthalpy (heat of the reaction) and $\Delta S d\xi$ is the change in entropy when the reaction progresses by $d\xi$. Let us consider two equilibrium states of the system that are infinitely close and separated by temperature and pressure intervals dT and dP. We can write that between these states the ΔG of the reaction remains unchanged:

$$d\Delta G = -\Delta S dT + \Delta V dP \tag{C.34}$$

giving Clapeyron's equation:

$$\frac{dP}{dT} = \frac{\Delta S}{\Delta V} = \frac{\Delta H}{T \Delta V} \tag{C.35}$$

where the second equation results from the condition that, at equilibrium, $\Delta G = 0$. This equation is extremely important in the Earth sciences as it is used for connecting the slope of phase change or mineralogical reaction in the pressure–temperature space familiar to geologists with two measurable quantities: the change in density (and so in molar volume) and the latent heat absorbed in the course of transformation.

The phase diagrams, which are representations of the stability field of the different phases that may appear within a system of given composition, are often approached qualitatively, but are also amenable to a thermodynamic approach. Let us consider a simplified granitic system, the quartz–albite (qz–ab) binary system, and ask which phases (quartz, albite, liquid) are stable at ambient pressure for a given proportion of albite/quartz and at a given temperature. If we consider that the latent heats of fusion of quartz and albite vary little with temperature, (C.32), applied to the solubility of quartz and albite, becomes after integration:

$$\ln x_{qz} = -\frac{\Delta H_{qz}(T, P)}{R} \left(\frac{1}{T} - \frac{1}{T_{qz}^f} \right) \tag{C.36}$$

$$\ln x_{ab} = -\frac{\Delta H_{ab}(T, P)}{R} \left(\frac{1}{T} - \frac{1}{T_{ab}^f} \right) \tag{C.37}$$

where the integration condition has been introduced so that, when the molar fraction x of each component is equal to unity, the temperature must be equal to the melting temperature T_{ab}^f and T_{qz}^f of the pure component. These two equations define the curve of the liquidus of the quartz–albite system (Fig. C.2). They cross at the eutectic point E. For each composition, the system has a stable branch at high temperature and a metastable branch at lower temperature.

If a liquid initially of composition x_A is cooled to temperature T_A, it allows some of the quartz to crystallize. Its composition evolves toward the eutectic point E where the albite begins to precipitate. If too little albite crystallizes, the system moves on to the metastable branch of quartz, on which it cannot remain without violating the minimum energy principle. If too much albite crystallizes, the system moves on to the metastable branch of albite, which is no more favorable either. The system's composition is therefore locked at the eutectic point (constant temperature and composition). When a solid system of the same composition melts, the pathway is reversed. The first liquids are of eutectic composition until the albite disappears. At that point the liquid evolves toward quartz and total melting is achieved when its composition returns to that of the initial

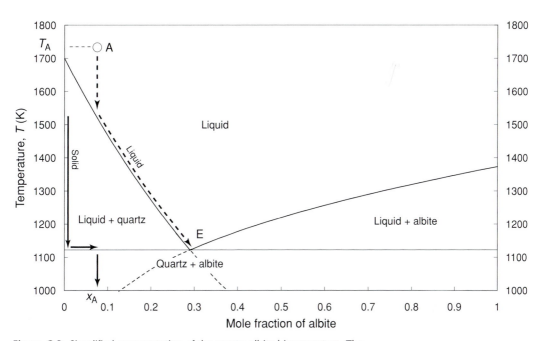

Figure C.2: Simplified representation of the quartz–albite binary system. The melting temperatures of quartz and albite are 1700 K and 1373 K, and their latent heats of fusion are 9400 and 63 000 J mol^{-1}, respectively. The liquidus curves are calculated from (C.36) and (C.37). The metastable part of the curves is shown as thin dashed lines. Their point of intersection is the eutectic point E, where a solid and a liquid of the same composition co-exist. Evolution by cooling of a liquid of composition x_A and initially at temperature T_A, and that of a solid at equilibrium are shown by the thick solid line (solid) and the thick dashed line (liquid). The path for melting would be the exact reverse.

solid. We have chosen a binary system in which solids do not form a mutual solution. Other systems whose components crystallize in similar forms (Fe–Mg olivine, Na–Ca feldspar) have solid solutions that may be treated in similar ways at the cost of slight complications in form. This approach is not specific to solid–liquid equilibria. It can be extended to phase equilibria of all sorts (solid–gas, solid–solid). Reaction phase equilibria are processed in the same way, which assemblages are termed peritectic assemblages.

Appendix D
The Rayleigh distillation equation

Let us consider a liquid system at high temperature and sufficiently well stirred for its composition to remain homogeneous. As the liquid cools, a solid phase appears and we are going to look at how the concentration of an element i changes as all of the liquid in the reservoir progressively crystallizes. We designate the properties of the newly crystallized solid by the subscript sol, those of the residual liquid by the subscript liq, and those of the liquid at the beginning of crystallization by the subscript 0. We write the mass of each phase (solid or liquid) as M and the mass of element i accommodated by one or other phase as m^i. We apply two conditions: that of the closed system (transformation of liquid to solid, without external inputs or outputs) and that of equilibrium fractionation of element i between the solid and liquid. The closed system condition is written in its incremental form:

$$\mathrm{d}M_{\mathrm{liq}} + \mathrm{d}M_{\mathrm{sol}} = 0 \tag{D.1}$$

$$\mathrm{d}m^i_{\mathrm{liq}} + \mathrm{d}m^i_{\mathrm{sol}} = 0 \tag{D.2}$$

The equilibrium condition between the last solid formed and the residual liquid is written using the partition coefficient $D^i_{\mathrm{s/l}}$ in the form:

$$\frac{\mathrm{d}m^i_{\mathrm{sol}}}{\mathrm{d}M_{\mathrm{sol}}} = D^i_{\mathrm{s/l}} \frac{m^i_{\mathrm{liq}}}{M_{\mathrm{liq}}} \tag{D.3}$$

Let us replace the properties of the liquid by those of the solid from (D.1) and (D.2):

$$\frac{\mathrm{d}m^i_{\mathrm{liq}}}{\mathrm{d}M_{\mathrm{liq}}} = D^i_{\mathrm{s/l}} \frac{m^i_{\mathrm{liq}}}{M_{\mathrm{liq}}} \tag{D.4}$$

or

$$\frac{dm^i_{liq}}{m^i_{liq}} = D^i_{s/l} \frac{dM_{liq}}{M_{liq}} \tag{D.5}$$

Taking the logarithm of concentration $C^i_{liq} = m^i_{liq}/M_{liq}$ of element i in the liquid and differentiating gives:

$$\frac{dC^i_{liq}}{C^i_{liq}} = \frac{dm^i_{liq}}{m^i_{liq}} - \frac{dM_{liq}}{M_{liq}} \tag{D.6}$$

By combining (D.5) and (D.6), and introducing the fraction of residual liquid $F = M_{liq}/M_0$, we obtain:

$$\frac{dC^i_{liq}}{C^i_{liq}} = (D^i_{s/l} - 1) \frac{dF}{F} \tag{D.7}$$

which is the form sought, applied to progressive crystallization of a liquid, and which leads us to (2.28).

Appendix E
The geological time scale

Era	Period		Epoch	Ended–began (Ma)
Cenozoic	Quaternary		Holocene	0–0.018
			Pleistocene	0.018–1.6
	Tertiary	Neogene	Pliocene	1.6–5.1
			Miocene	5.1–24
		Paleogene	Oligocene	24–38
			Eocene	38–55
			Paleocene	55–65
Mesozoic	Cretaceous			65–144
	Jurassic			144–213
	Triassic			213–248
Paleozoic	Permian			248–286
	Carboniferous			286–360
	Devonian			360–408
	Silurian			408–438
	Ordovician			438–505
	Cambrian			505–570
Proterozoic				570–2700
Archean				2700–4000
Hadean				4000–4560

Appendix F
An overview of analytical methods

The analytical methods of geochemistry are many and varied, but they can be grouped by family depending on what is to be analyzed. Setting aside the high-temperature and high-pressure experiments that involve methods often borrowed from mineralogy and petrology, these methods fall roughly into three groups depending on what is to be measured:

- concentrations;
- isotopic ratios;
- speciation of elements in solutions, mineral phases, or organic matter.

We will omit the last item here, as the variety of methods would involve substantial developments with a large physics content about spectroscopic methods.

Measurement of concentrations

Two general principles are commonly used. The first one uses comparison with a reference material by means of calibration curves and only requires from-the-shelf reagents, while the second one, isotope dilution, requires artificially altered nuclide mixtures. In the first case, the operator compares the response to physical stimulation (radiation, ionization) by means of a suitable detector upon the passage of a solution containing the dissolved sample and a set of reference solutions. The first step of most procedures is the dissolution of the powdered sample in hydrofluoric acid (HF), the only acid to dissolve silicates. Often this attack phase is replaced by melting of the sample powder in a lithium meta-borate "flux," the addition of which lowers the melting point of the mixture for all the minerals, even the most refractory ones (zircon, oxides). The resulting glass can be dissolved in hydrochloric acid, which is far less dangerous than HF. The attack solution is then diluted so as to minimize problems of interference between the elements and the solution (matrix effects). Let us review four types of methods, each being associated with a different detector:

1. *Atomic emission spectroscopy* (AES) operates by light excitation of rays specific to an element. The method involves spraying the solution in a flame and is excellent for measuring the alkaline elements. As in all optical methods, the different wavelengths of the light beam are separated by an optical grating, a device which sorts the wavelengths by their interference patterns. Inductive plasma excitation, which we will return to later, excites virtually all the elements. The large number of energy levels per element makes it easier to select rays with no interference with the matrix elements. The method is therefore relatively sensitive, but the interference problems between elements are complex.

2. *Atomic absorption spectroscopy* (AAS) relies on the opposite effect. Monochromatic light sources illuminate a vapor produced by heating the solutions in a graphite oven so as to excite electrons to higher energy levels. Excitation is particularly effective when the incident radiation energy corresponds to the transition energy (resonance). For a given element, the relative absorption energies of the sample and the standards are used to determine concentrations.

3. *X-ray fluorescence spectroscopy* (XRF) relies on the exposure of the sample powder to X-rays. The sample re-emits X-rays of lower energy by fluorescence and their wavelengths are analyzed. Each element appears as a peak whose intensity can be calibrated against standards. Measurements are precise, but below 100 ppm the method is not sensitive enough to produce good results.

4. *Inductively-coupled plasma mass-spectrometry* (ICP-MS) relies on ionizing the solutions to be analyzed by inductive coupling of the solution nebulized in a stream of argon within a plasma torch. The ICP acronym, standing for inductively coupled plasma, is a buzzword that the would-be geochemist must remember. Argon is chosen for its high-ionization energy: once ionized as Ar^+ by induction in a coil, it captures the least firmly bonded electron of the other ions vaporized in the plasma. The different masses are separated in the ion beam by a simple, lightweight mass spectrometer known as a quadrupole: the trajectory of ions injected between four bars with crossed potential is disturbed by radio frequencies so that only the ions of a specific mass cross the device and are counted by the detector. This type of device has no magnets and so allows very fast scanning across the mass range. Masses close to those of Ar, ArO, ArN (40–56) or to the atmospheric gases (N, O, C) cannot be used, as the detector is saturated by the gas stream. For other masses, the respective heights of the signals are compared for the sample and the standard solutions. This method is both very sensitive

(detection level of less than 10^{-11} g $= 10$ pg), and very precise (1–2%). Even so, it yields only poor isotopic ratios (at best 2%), as the tops of the mass peaks are not flat and the counting statistics marginal (see below). As an alternative to dissolution, the rock or mineral sample may be sprayed into the argon stream with a laser beam, preferably operated in the ultraviolet wavelength range. The remainder of the protocol (ionization in the plasma torch and mass analysis) remains unchanged. This method, known as laser ablation, allows *in situ* analysis of concentrations with a beam size of less than 30 micrometers and a remarkably low detection level.

Isotope dilution is a complex, but exact, reference method operating on a different principle; the equations will not be set out in detail here. The element in question must have more than one isotope, which, in particular, excludes phosphorus, aluminum, manganese, cobalt, etc. To cut a long story short, let us use another animal analogy and take as our job counting goldfish in a pond. We take some goldfish of the same species from a neighboring pond, count them and mark them, and then put the marked goldfish into the pond we are interested in. After a few days, so that the marked fish are evenly distributed in the pond, a net is thrown in and the proportion of marked goldfish relative to the unmarked ones is rapidly determined before returning the fish to the water. Simply dividing the number of marked fish introduced by the proportion of marked fish collected gives the total number of fish and therefore the number of fish initially present in the pond. The marked fish are termed the spike. If ^{85}Rb and ^{87}Rb are substituted for the marked and unmarked fish, we have the measurement of rubidium concentrations by isotopic dilution. By measuring the proportion of each isotope, we gain access to the number of atoms of each of them initially present. Isotopic spikes are produced by specialist firms. A remarkable property of this method is that the size of the sample collected by the net does not affect the result (provided a minimum number of individuals is taken): once the spike is added and has completely mixed with the sample, the mixing ratio can be obtained from any fraction of the mixture, however incomplete. Isotope dilution does not require a full recovery of the element to be analyzed. This absolute character of the method, offset unfortunately by its painstaking nature, makes it an outstanding reference method.

Measurement of isotope compositions

This measurement principle is very different from those governing the methods described above, as we seek to measure the ratio of two signals emitted simultaneously corresponding to two different isotopes of the same element. Because the ratio of two simultaneous signals is

Figure F.1: Mass
spectrometer. A source emits
ions that are accelerated by a
high voltage and then
separated in a magnetic field
according to their mass (M_1,
M_2, etc.). Ion beams are
collected in Faraday cages,
converted to voltages by a
high-value resistor, and these
voltages, which are
proportional to isotopic
abundances, are analyzed by
an array of voltmeters.

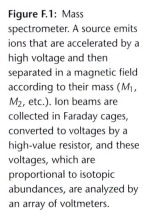
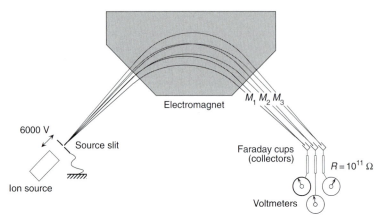

compared, the precision of isotopic measurements is far better than that
of elemental analysis, for which the intensity of the signal itself is the
measure. Figure F.1 is a diagram of a mass spectrometer. The different
ion sources described in the following correspond to different physical
processes and to different applications:

1. *Gas bombardment* sources are based on the principle of spraying the sample
    gas with electrons, which strip a peripheral electron from it. This type of
    source is used for measuring the isotopic compositions of hydrogen (H_2 gas),
    carbon and oxygen (CO_2 gas), sulfur (SO_2 gas), and also the inert gases (He,
    Ne, Ar, Xe). The use of lasers for *in situ* analysis of isotopic compositions, in
    particular of oxygen in minerals and of argon for geochronology, now allows
    the operator to select the domains to be analyzed so as to avoid mineral rims
    and altered material.
2. *Thermal ionization* (TIMS) sources are based on the probability of ions
    rather than neutral atoms being ejected from the surface of a sample placed
    on a filament – usually rhenium, tantalum, or tungsten – and heated to
    a high temperature (1200–1800 °C) in a vacuum. This method has dom-
    inated isotopic measurement of positive ions of Sr, Nd, Pb and Th, and
    negative ions of Os for decades. However, many elements, specially those
    with high ionization potential, cannot be analyzed this way. In addition,
    the isotopic compositions of elements such as lead that do not have at
    least two stable isotopes cannot be corrected precisely for analytic mass
    bias.
3. The *inductively-coupled plasma* (ICP) sources have been described above.
    These most successful instruments couple this type of source with a magnetic
    sector and an array of multiple collectors (MC-ICP-MS). For many elements,
    this method, which emerged in the mid-1990s, met with considerable success
    for two main reasons: the capacity of the plasma torch to ionize all elements,
    including those too refractory to be analyzed by TIMS, and the possibility
    of correcting analytical bias precisely for all elements. The results of this
    technique for radiogenic isotopes Hf, Nd, and Pb, but also for the "new"

stable isotopes (B, Fe, Cu, Zn) make this method the successor to thermal ionization.

4. *Secondary-ion sputtering* (SIMS), known as the ion probe, uses a primary beam of ions, typically negative oxygen ions or positive cesium ions, to spray the sample surface and sputter the sample ions into the instrument. This very sensitive method is well adapted to *in situ* isotope measurement and is particularly successful for *in situ* U–Pb dating of zircons (especially the famous SHRIMP of Canberra, Australia). It is poised to become the prime method for *in situ* analysis of isotopic compositions of carbon or oxygen.

Once the atoms from the sample are ionized, the ion beam is then accelerated at several thousand volts and spread in mass in a magnetic prism. The beam corresponding to each mass is normally collected in a Faraday cage, which often is a carbon-coated ceramic box attached to a high-value resistor, typically $R = 10^{11}\ \Omega$ (ohms), and a voltmeter. By Ohm's law, the reading of the flux of ionic charges I, i.e. the electric current, can be replaced by the reading of the potential difference $V = RI$, which is usually of the order of one volt. When the ion beam is too weak, the analog current reading is replaced by the counting of the individual arrivals of charged ions. Modern techniques allow this to be read simultaneously from several beams corresponding to different masses (multiple collection) and so make for more precise measurement.

To measure isotopic ratios of interest for radiogenic isotopes, e.g. the $^{87}\mathrm{Sr}/^{86}\mathrm{Sr}$ ratio, both isotopic instrument bias and natural thermodynamic and kinetic fractionation are eliminated by standardization to an arbitrary stable isotope ratio (Fig. F.2). For strontium, the universal choice is $^{86}\mathrm{Sr}/^{88}\mathrm{Sr} = 0.1194$. From a linear approximation (2.23), the magnitude f of the mass bias can be inferred from the measurement of this ratio:

$$\left(\frac{^{86}\mathrm{Sr}}{^{88}\mathrm{Sr}}\right)_{measured} = 0.1194\,(1 - 2f) \tag{F.1}$$

and the resulting value of f introduced into the similar expression for the $^{87}\mathrm{Sr}/^{86}\mathrm{Sr}$ ratio:

$$\left(\frac{^{87}\mathrm{Sr}}{^{86}\mathrm{Sr}}\right)_{normalized} = \left(\frac{^{87}\mathrm{Sr}}{^{86}\mathrm{Sr}}\right)_{measured} \frac{1}{1 + 1f} \tag{F.2}$$

Note the values -2 and $+1$ for the mass differences, between the masses at the numerator and denominator of the isotopic ratios. Agreement is not complete for standardization of isotopic ratios of certain elements such as neodymium and values of the $^{143}\mathrm{Nd}/^{144}\mathrm{Nd}$ ratios taken from the literature must be used with care. This problem of calibration of bias among laboratories disappears when reference notation to a common

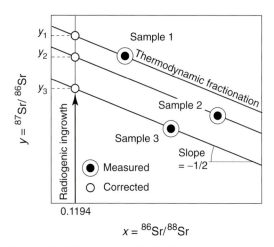

Figure F.2: Principle of standardization of radiogenic isotope abundances (here $y = {}^{87}Sr/{}^{86}Sr$) relative to a reference ratio (here $x = {}^{86}Sr/{}^{88}Sr$). The various straight lines represent both natural and analytic thermodynamic isotopic fractionation depending on mass. The different ordinates between the straight lines represent the effect of radioactive accumulation of ${}^{87}Sr$, which varies from one sample to another. Standardization to $x = 0.1194$ eliminates thermodynamic fractionation leaving only radiogenic variability y_1, y_2, and y_3. Note the slope $-1/2$ equal to the $(87\text{--}86)/(86\text{--}88)$ ratio of mass differences.

standard sample is used (delta or epsilon notation). Lead has only a single stable isotope, which rules out internal standardization and explains the intrinsically lower precision of measurement of its isotopic compositions (a few per mil) compared with that of Sr or Nd ($10 - 30 \times 10^{-6}$).

Two remarks are called for so as to judge the relevance and quality of a method. First, for all the methods discussed, atoms are in competition for ionization. A mixture of atoms practically never gives an acceptable ionization yield. Except for ionic analysis, which does not allow it, we seek to isolate the element to be analyzed from the other components (matrix), while taking care to maintain a very high chemical separation yield. This is why complex separation procedures must be carried out (extraction lines for gases and stable isotopes, ion exchange on resins in solution for thermal ionization and plasma sources) before performing mass spectrometry.

This brings us to the second point, which is the requirement of precision. While a precision of 1% can give good-quality information for the isotopic analysis of argon, precision better than 10^{-3} is required for meaningful lead and osmium isotopic data, and better than 10^{-4} for Hf, Nd, and Sr. How do we judge the required precision? A counting process is a Poisson process, i.e. the probability of an event (arrival of an ion) occurring depends only on the duration of measurement and not on the moment at which it is made. For this type of process, the standard deviation of measurement (which yields the precision) is equal to the square root of the number of events. If, say, we count a million ions, the standard deviation of the measurement will be 1000 counts and the relative precision of counting therefore will be one per 1000. Understandably then, this precision depends mainly on the number of ions analyzed. The flux of particles cannot be increased indefinitely as the ion, electron, and photon counting devices lose their linearity for a few hundred thousand strikes per second and are totally saturated at a few million. For isotopic composition of lead, a significant measurement can be obtained on an impact of a few tens of micrometers, on ionic analyzers such as the SHRIMP working at the extreme limit of experimental possibilities. Given the current state of sensitivity of instruments, it would be pointless, for example, to hope for a significant measurement on the same impact of the isotopic composition of Nd, as the quantity of the element required would be more than a hundred times greater than that required for lead. The elementary processes of physics must be comprehended and measurement can only be refined by improving the efficiency of ionization.

Appendix G
Physical and geophysical constants

Physical constants	
Avogadro number, $\mathcal{N}$ (mol^{-1})	$6.022\,137 \times 10^{23}$
Gas constant, R (J mol^{-1} K^{-1})	$8.314\,51$
Mean depth (km)	
Earth center	6371
Solid core	5150
Core–mantle boundary	2891
Upper mantle	660
Oceanic lithosphere	80
Oceanic crust	7.1
Continental crust	39
Ocean	3.77
Continental elevation	0.825
Surface areas (m^2)	
Whole Earth	5.10×10^{14}
Land	1.48×10^{14}
Masses (kg)	
Whole Earth	5.9736×10^{24}
Inner core	9.84×10^{22}
Outer core	1.841×10^{24}
Mantle	4.002×10^{24}
Lower mantle	2.940×10^{24}
Upper mantle	1.062×10^{24}
Continental crust	2.97×10^{22}
Oceans	1.40×10^{21}

Polar ice and glaciers	4.34×10^{19}
Atmosphere	5.1×10^{18}
Mean densities (kg m^{-3})	
Whole Earth	5526
Core	10 957
Lower mantle	4903
Upper mantle	3604
Fluxes	
Oceanic crust at ridge crests (km^2 a^{-1})	3.5
Rivers and runoff (kg a^{-1})	3.6×10^{16}
Mechanical erosion (kg a^{-1})	2.0×10^{13}

Appendix H
Some equations relative to residence time

First proposition: The relaxation time τ^i is the mean residence time of element i in the reservoir. For an upstream concentration of zero, this essential property can be demonstrated from (5.2) as follows. The mean time $\langle t \rangle$ that element i spends in the reservoir is the mean algebraic value of variable t summed for all increments of mass $dM^i(t)$ of the element passing through the reservoir between $t = 0$ and $t = \infty$:

$$\langle t \rangle = \frac{1}{MC^i(\infty) - MC^i(0)} \int_{MC^i(0)}^{MC^i(\infty)} t \, d\left[MC^i(t)\right] \tag{H.1}$$

where the masses of element i contained in the reservoir at those times are given by simply multiplying the concentrations by the mass M of the reservoir. Under the chosen conditions, concentration $C^i(\infty)$ is zero, mass M is constant and cancels out. By taking the derivative of $C^i(t)$ obtained from (5.3), and by introducing the variable $u = t/\tau^i$, we obtain:

$$\langle t \rangle = \tau^i \int_0^\infty u e^{-u} du \tag{H.2}$$

Integrating by parts, i.e. using:

$$\int u e^{-u} du = -u e^{-u} + \int e^{-u} du \tag{H.3}$$

shows that the sum in (H.2) equals unity and demonstrates the proposition. A similar demonstration applies for non-zero upstream concentrations. Observe that if the mass of the reservoir varies with time, the demonstration is no longer valid.

Second proposition: Extend (5.3) to the case of a constant upstream concentration. When the concentration C^i_{in} is constant, (5.2) can be rewritten as:

$$\tau^i \frac{d\left[C^i - C^i_{in}/\left(1 + \beta^i\right)\right]}{dt} = -\left(C^i - \frac{C^i_{in}}{1 + \beta^i}\right) \tag{H.4}$$

a homogeneous first-order equation that integrates simply as:

$$\left(C^i - \frac{C^i_{in}}{1 + \beta^i}\right)_t = \left(C^i - \frac{C^i_{in}}{1 + \beta^i}\right)_0 e^{-t/\tau^i} \tag{H.5}$$

or

$$C^i(t) = C^i(0)e^{-t/\tau^i} + \frac{C^i_{in}}{1 + \beta^i}\left(1 - e^{-t/\tau^i}\right) \tag{H.6}$$

where the right-hand side contains the sum of the relaxation term and the forcing term.

Third proposition: In a well-mixed system at steady state, the distribution of residence times (the histogram of age groups) is exponential. Let us call $n(t, \theta)$ the number of particles that, at time t, have resided in the system for time θ. Let us call $p = 1/\tau$ the probability that a particle leaves the system in a unit time. The assumption that the system is well mixed amounts to a constant p, independent of θ. We will illustrate the "Lagrangian" point of view of this problem with a human population comparison. Let us consider in a human population at time t the group of 65-year-old individuals. Each year, a fresh group is formed from all the 64-year olds, while all 65-year olds will either be promoted to the older age group or die. In our geological system, the number $n(t, \theta)$ of particles changes because particles exit the system but also because particles age. The rate of change is given by:

$$dn(t, \theta) = -pn(t, \theta)\,dt - n(t, \theta + d\theta) + n(t, \theta) \tag{H.7}$$

where:

$$n(t, \theta + d\theta) = n(t, \theta) + \frac{\partial n(t, \theta)}{\partial \theta}d\theta \tag{H.8}$$

Every second, the residence time of a particle that remains in the system increases by one second and, therefore, $d\theta = dt$. We can rearrange the previous equation as:

$$\frac{\partial n(t, \theta)}{\partial t} = -pn(t, \theta)\,dt - \frac{\partial n(t, \theta)}{\partial \theta} \tag{H.9}$$

At steady state, the left-hand side of this equation vanishes and the right-hand side may be integrated as:

$$n(t, \theta) = n(t, 0)\, e^{-p\theta} = n(t, 0)\, e^{-\theta/\tau} \qquad\qquad \text{(H.10)}$$

which demonstrates our second proposition. By integrating this expression over the entire population, we would find that the mean residence time is simply τ.

Further reading

Level of difficulty: (A) armchair reading; (B) for students; (C) serious stuff. This list of references is not exhaustive. It is only intended to guide the reader into more specialized fields.

Albarède, F. (1995) *Introduction to Geochemical Modeling.* Cambridge, Cambridge University Press. Geochemical modeling methods with plenty of examples. C.

Allègre, C. J. (1992) *From Stone to Star.* Harvard, Harvard University Press. The history of Earth Science concepts by one of geochemistry's main players. A.

Anderson, D. L. (1989) *Theory of the Earth.* Oxford, Blackwell. The original ideas of one of the most influential geophysicists who does not hesitate to cross the line between geophysics and geochemistry. B.

Aris, R. (1999) *Elementary Chemical Reactor Analysis.* Mineola, Dover. A reprint of the 1989 classic textbook on chemical engineering. Outstanding value. C.

Berner, R. A. (1980) *Early Diagenesis.* Princeton, Princeton University Press. A common sense but powerful explanation of early diagenesis. B.

Broecker, W. S. (1994) *Greenhouse Puzzles.* Palisades, Eldigio. The greenhouse effect will never seem the same to you again. A.

Broecker, W. S. (1995) *The Glacial World According to Wally.* Palisades, Eldigio. Quaternary climates by the undisputed champion. A.

Broecker, W. S. and Peng, T. H. (1982) *Tracers in the Sea.* Palisades, Eldigio. Not read it yet? Still fascinating 20 years later. B.

Brownlow, A. H. (1986) *Geochemistry.* Upper Saddle Review, Prentice Hall. One of the most popular textbooks on geochemistry. B.

Condie, K. C. (1986) *Mantle Plumes and Their Record in Earth History.* Cambridge, Cambridge University Press. Everything you could want to know about plumes in the geological record. C.

Criss, R. E. (1999) *Principles of Stable Isotope Distribution.* Oxford, Oxford University Press. An enjoyable reference with examples. C.

Davies, G. F. (1999) *Dynamic Earth: Plates, Plumes, and Mantle Convection.* Cambridge, Cambridge University Press. Simple but intelligent physics reveals a lot about the Earth's interior. A must. B.

Denbigh, K. (1981) *The Principles of Chemical Equilibrium.* Cambridge, Cambridge University Press. Many have tried to do better. I keep patching up the cover of my copy. B.

Dickin, A. (1995) *Isotope Geochemistry.* Cambridge, Cambridge University Press. An important reference work in radiogenic isotope geochemistry. B.

Faure, G. (1986) *Principles of Isotope Geochemistry*. Chichester, Wiley. A classic that is withstanding the erosion of time. B.

Greenwood, N. N. and Earnshaw, A. (1995) *Chemistry of the Elements*. Oxford, Butterworth-Heinemann. A best-seller on the inorganic chemistry of the elements.

Hinchliffe, A. (1999) *Chemical Modeling from Atoms to Liquids*. Chichester, Wiley. A remarkably insightful and readable introduction to a reputedly difficult field. B.

Karplus, M. and Porter, R. N. (1970) *Atoms and Molecules*. Benjamin. For an intimate understanding of matter. C.

Kondepudi, D. and Prigogine, I. (1998) *Modern Thermodynamics: From Heat Engines to Dissipative Structures*. Chichester, Wiley. From equilibrium to irreversible processes: particularly broad, insightful, and enjoyable. B.

Krauskopf, K. B. and Bird, D. K. (1995) *Geochemistry*. New York, McGraw-Hill. One of the most popular textbooks on geochemistry. The third edition was extensively rewritten. B.

Lasaga, A. (1998) *Kinetic Theory in the Earth Sciences*. Princeton, Princeton University Press. An outstanding account of the physics that underlies geochemistry. C.

Lerman, A. (1988) *Geochemical Processes*. Malabar, Krieger. An amazing book on low-temperature geochemistry with a unique approach to physical processes. Never out of fashion. B.

Maczek, A. (1998) *Statistical Mechanics*. Oxford, Oxford University Press. A bargain book that leads the reader swiftly through the arcanes of elemental and isotopic fractionation. B.

McBirney, A. R. (1992) *Igneous Petrology*. San Francisco, Freeman-Cooper. An elegant and non-conventional introduction to magmatic processes. B.

McSween, H. Y., Jr (1999) *Meteorites and Their Parent Planets*. Cambridge, Cambridge University Press. An excellent modern textbook on these weird and fascinating objects. B.

Millero, F. J. (1996) *Chemical Oceanography*. Boca Raton, CRC Press. An authoritative reference. B.

Morel, F. M. M. and Hering, J. G. (1993) *Principles and Applications of Aquatic Chemistry*. Chichester, Wiley. A particularly insightful introduction to usually difficult concepts of water chemistry. C.

The Oceanography Course Team of the Open University (1991) Oxford, Pergamon. These seven titles of very clear and wonderfully illustrated bargain books supersede any introductory text in Oceanography. I recommend particularly two of these: *Ocean Circulation* and *Seawater: Its Composition, Properties, and Behavior*. A.

Ottonello, G. (2000) *Principles of Geochemistry*. New York, Columbia. An excellent textbook focused on thermodynamics, now in an affordable paperback edition. B.

Richardson, S. M. and McSween, H. Y., Jr (1989) *Geochemistry: Pathways and Processes*. Upper Saddle River, Prentice Hall. A balanced view of geochemistry. B.

Stumm, W. and Morgan, J. J. (1995) *Aquatic Chemistry*. Chichester, Wiley. The undisputed bible with plentiful examples. C.

Taylor, S. R. (2001) *Solar System Evolution: A New Perspective*. Cambridge, Cambridge University Press. An extremely well-written textbook by one of the leading scholars of geochemistry. B.

Taylor, S. R. and McLennan, S. M. (1985) *The Continental Crust: its Composition and Evolution*. Oxford, Blackwell. The ultimate read on the geochemistry of continental crust. B.

Valley, J. W. and Cole, D. R. (2001) *Stable Isotope Geochemistry.* Washington, Mineralogical Society of America. A set of varied and very high-quality review papers. B.

Index